More praise for *Otter Country*:

'Beautifully evocative . . . Written in prose as sinuous as the creatures themselves . . . Otters apparently enjoy a rather illustrious club of fans. Readers of Ms Darlington will be glad she is among them' *Economist*

'Truly lyrical . . . wonderfully written' *Independent*

'A diary of utter devotion where landscape is interpreted as otter-text, its messages elusive but traced with deep sensitivity' Jay Griffiths, author of *Wild: An Elemental Journey*

'A truly mesmerising and inspirational read' *Countryside*

'Darlington is consistently an agreeable companion – inquisitive, but never a know-it-all, and frequently funny . . . I challenge anyone not to find [her] writing buoyant and inspirational' *Country Life*

'A passionate journey, travelled beautifully' *BBC Wildlife Magazine*

'The writing is sublime . . . a piece of sheer joy' *West Country*

'A book to restore that sense of nature red in tooth and claw and our place in amongst it . . . I turned the last page of *Otter Country* full of admiration' Dove Grey Reader

'The known and loved landscapes of her past and present are evoked with as much care as she gives to her musteline quarry' Esther Woolfson, *Literary Review*

'Wonderful . . . her book is equal to its subject' *Financial Times*

'Sharply observed and lyrically written' *Spectator*

OTTER COUNTRY

In Search of the Wild Otter

Miriam Darlington

Illustrated by Kelly Dyson

GRANTA

Granta Publications, 12 Addison Avenue, London W11 4QR

First published in Great Britain by Granta Books 2012
This paperback edition published by Granta Books 2013

A CIP catalogue record for this book is available from the British Library.

1 3 5 7 9 10 8 6 4 2

ISBN 978 1 84708 486 6

Typeset by Avon DataSet Ltd, Bidford on Avon, Warwickshire

Printed and bound in Great Britain by CPI Group (UK) Ltd, Croydon, CR0 4YY

For my family

CONTENTS

Spirit Level   1

True North   15

Watershed   65

Marsh   113

Source   155

Stream   203

Hunting Ground   239

East   277

Holt   309

Acknowledgements   341

Permissions   345

Bibliography and Further Reading   348

Index   354

Spirit Level

I dream of them,
crossing like pirates the East Anglian tundra,
ears honed to the wind,

carved out of laughter and let loose
more currents and bends in their bodies
than a whole river.

They mean we are still alive,
that land's edges have not lost entirely
their mapless unknown.

Kenneth Steven, 'Otters'

It's midsummer and I want to go north. Somewhere it'll be hard for anyone to find me. Away from the hubbub of signposts, maps and people. I want the wild. Some edge where daylight stays up late into the night. I want antlers of deer silhouetted against luminous sky. I want wildcats, peat bog, Caledonian forest.

I pack my camper van with waterproofs, a notebook and a battered copy of *Ring of Bright Water*. Images from Maxwell's classic tale about living with otters have been imprinted in my mind ever since I read it as a child. Never mind the slippery reality it portrays, the spell of its otters took hold of me long ago and I still long to see one in the wild.

After hours of travelling, the dazzle of tarmac and white lines blurs my senses. I pull into a lay-by in the moonscape of Rannoch Moor. The panorama of water, rock and heather seems like a gateway to another world. In this wide expanse of peaks and distances anything human is dwarfed. The ground reveals half-digested stone remains, hinting at its history of struggle and clearance. A short distance away, pools in the peat create mirages of solid earth where living creatures

3

can be swallowed. Deeper down, older shadows lie hidden. The bog has been nourishing itself for millennia, and beneath the visible surface is the dust of mammoth, elk, wolf and bear. If I walked away from the metallic road and into the ghostliness of it all, I might easily disappear.

Away from the driving seat I can hear, somewhere high up, a skylark chipping away at higher and higher notes. I step off the grit of the road and move into a live surface of sphagnum moss. Close-to, the silent, mottled peat bog comes into focus, forming a mapless forest of colour. Through it all a vehicle rages by at what seems like a crazy speed, ripping the air with a wake of noise.

Soft rush and molinia grass offer resistance against my boots. The air is saturated with scents of animal, peat and gorse. Each day this moist ground spawns a million midges. *Culicoides impunctatus*; their intricate mandibles are designed to pierce the skin and suck a meal from any available mammal. Water surfaces sensitive to the slightest touch are sprinkled with minute imprints that could be particles of mizzle, or micro-feet and feelers, all ready to lift off. I get back into the van, open the windows and drive more slowly. Other traffic has dropped away, and the Munros of Wester Ross rise from the dwindling road. Further on, I enter the slopes and peaks of Torridon. Here, remnants of wildness hit me where the ribs meet and leave no room to breathe. Still water on the sea loch glimmers with the jagged presence of something immense. In my visitor's guide I read that Torridon means 'heather' in Gaelic. But it has also been translated as 'place of transference'. In my road-sore and eye-weary state, the word 'transference' sounds as if those who pass through will be subject to some kind of metaphysical change.

Towards Gairloch the earth begins to smooth out into green folds of machair. It rumples into dunes, or crinkles like scaled skin down to the tangle of the inter-tidal zone, the weave in the border between water and land. The road map bears no relation to the waves, salt-wind and ruggedness here; approaching Little Loch Broom, I throw it into the back, cut the engine and roll as close to the sea as I dare. As I step out, the tang of bladderwrack and kelp fills my nose. High above, coruscating peaks form a skyline of names that sound like broken teeth: Beinn Dearg, An Teallach, Beinn Ghobhlach. It feels bizarre to be alone amongst all this magnificence. But at the same time, it's essential.

My body is churning with the journey, with the trance of tarmac unspooling into sheep, red deer, tumbling streams. Unable to take it all in, I get back into the van, remove my boots and socks, and stretch out on my bunk. When I try to sleep, surfaces of rock and shivering thrift rise behind my eyelids as if the landscape is already dreaming its way into me. I lie listening to the tick of the engine as it cools, and some time later my sleep is broken by a new sound. I surface and fumble for my watch. It's broad daylight, but not yet 4 a.m. The air is saturated with the smell of wet seaweed. And there it is again: a cracking sound, magnified by the moisture on the air.

I peer out, expecting to see a winged creature, a fisherman or something with hooves startling away into the mist. I struggle to focus. Then I see it. An otter. Three or four metres away, hunched low on a flat rock. Just over a metre in length, he has the dimensions of a male or dog otter, with a broad, flat head, large back feet and a long, tapering tail. It's the magnificent ruff of whiskers that surprises me, and the bulk of him, the fur sleek from fishing out in the loch. I am seized with joy; his eyes look through me as if I don't exist.

I watch, puffy-eyed, nostrils pressed to the glass, as the otter devours his meal. He seems to have tackled the pincers first, which makes sense. Using delicate webbed fingers that are almost human in their dexterity, he doesn't waste a moment. He holds down his prey and crunches through claws, shell and flesh. Turning his muzzle up to the sky, he swallows, jaws powerful as a trap, spreading splinters all around. His long body is smooth as a lithe brown muscle. I marvel at his ability to make a meal out of an entire crustacean, as every last piece disappears into the pink of his gullet.

As if in an elaborate ritual, the otter begins to preen his wet pelt in preparation for sleep. What was smooth as brown leather has become spiked as a hedgehog, lightening in colour as he dries. He moves to the grass, selects his spot and rolls sinuously over and over, exposing a silvery throat and milk chocolate beneath to the sky. Supple as rope made out of silk, he nuzzles his flanks and like a cat he wriggles and rolls in the sun, then gradually coils himself into a pretzel of fur. I watch the rise and fall of the otter's sleeping form. His pelt has lightened to chestnut brown now he's dry, and he is knotted so tightly that limbs, head and tail are indistinguishable. Still in last night's clothes, I slip out and try to get a little closer to the curled shape in the grass. My stealthy commando crawl drags in the heather, making what I think is only the subtlest of sounds.

The otter seemed deeply asleep but one ear must have been left on duty, because soon his head rises. Instantly he's alert. His neck lengthens for a moment and his head bobs as he makes sense of the threat. I can tell he is very puzzled; maybe he thinks I am a bigger otter? From this sniff-level position I get a flash of the bristling vibrissae, the otter's extravagant whiskers, and in a split second he catches my scent. He turns and makes a direct gallop for the shoreline. He moves quickly, but with the lumbering gait that, I learn later, otters always have on land. His body arches into a small hummock as he runs, but when he disappears back into the water, he does so without a sound. In the stillness of the June dawn there is hardly a breath of wind, and the last thing I see is the tip of his tail as it's sucked into the glassy sea. The tussock grasses are still, the white tufts of cotton immobile and the otter dissolved into the water as if he were part of it, his colour and shape sleeking into nothing.

The air sifts through the grass. An oystercatcher opens its thin red bill and calls in alarm. A freshwater stream dribbles over pebbles nearby; I can smell sheep fleece, layers of wet peat and sweet bog myrtle. Tufts of sea pink flutter as if the whole earth is flowing with electric current. I want to go after my otter, but he's already melted away.

When I return home to Devon, I'm restless and can't settle. I remember another pivotal moment, when I was just eleven. I was given an otter skull to hold. The strangeness of cool bone filled my palm; sharp canines, flattened crown, broad eye sockets. It was more streamlined than any other skull I'd seen. The hinge of the jaw opened and snapped shut on a perfect jigsaw of incisors. I turned it over and over. Here was a three-dimensional map of the evolution of a predator: large canines for gripping fish, shearing carnassials, strong shell-grinding teeth for dealing with harder prey. I knew about predators in far-away places: tigers, lions, alligators, wolves. But this one was local, and its skull fitted into my hand.

It was normal in my family to be given this sort of object. The house of my grandfather, the biologist C.D. Darlington, was stacked with his books on genetics and evolution. His shelves were a professorial museum of fossils and artefacts, and he seemed as old as many of them. When he answered my questions, he made all these bones seem alive with resonances. Charles Darwin said: 'How paramount the future is to the present when one is surrounded by children.' Perhaps my grandpa was thinking this as he spoke to me. After he died, a whole web of connections slowly came into focus. Ideas began to grow in my mind: fossils were not just dead things; bones had ancestors, families, loved ones.

Surrounded by an inherited hoard of skeletal curiosities, my mother attempted to distract me with poetry. I found Ted Hughes's poem about an otter, where the hunted animal is reduced to nothing but a skin over the back of a chair. I knew that the otter's future was precarious and I worried that there might already be no otters left where I lived.

I went to my local library and found the books on carnivores; I made a pile on the table and searched through the pages until I found an otter. I imagined that the sour dusty ink-scent of the page smelled like otter. I stared and stared at the pictures of this elusive animal. Then I began reading: the otter, I discovered, was one of the earliest mammalian carnivores to evolve. Its ancestral genus, *Mionictis*, occurred about thirty million years ago. Scientists estimate that we branched off from our common ancestor with chimpanzees between five and seven million years ago. The otter had been here long before us. And well-preserved otter-like skeletons were still being discovered, each one hinting at a little more information: how these mammals may have adapted, coming to lake shores and rivers where they found new feeding at the water's edge, the flattened toes in some skeletons indicating early webbed feet.

I persuaded my parents to take me to the Booth Museum, near where we lived in Sussex. The museum was built in 1874 to house a million pinioned butterflies, hundreds of stuffed birds and every skeleton imaginable – an Aladdin's cave, a hoard of obsessive, grisly Victorian science. In the absence of a real live otter, I could stare through the glass at the next best thing: an *entire* otter skeleton. The small, smooth cranium gleamed spookily in the soft lighting. I examined the blunt snout and intricate, concave nose cavity designed for heightened scent-capture; the impressive sharp-edged teeth set in a compact but powerful jaw. The supple spine flowed into an ottery hump, typical, apparently, of its shape when moving on land. Strong vertebrae held the large ribcage of a chest adapted for underwater survival; all of it tapering into an exuberantly long, rudder-like tail. The otter's short legs and elongated feet, with five delicately webbed toes, would help it to accelerate under water or forage on land. Streamlined and flexible, it could also move easily over marsh or silt without sinking. Positioned into a semi-natural pose in its case, I could imagine it glaring at its captors, racing after fish underwater, or nuzzling its brood of cubs in a den by some riverbank. Inside those bones was a story so large it stretched my mind and called up thoughts that left me dizzy. Too close for comfort, other skeletons haunted their own glass cases: the Dodo, the Great Auk, the Tasmanian Tiger, all marked *Extinct*.

Shortly after my encounter with the otter skull, I read Henry Williamson's *Tarka the Otter*. The life story of this fictional otter had some kind of alchemical effect on me. For months afterwards I felt like I was an otter, as if the story had created some kind of kinship between this small predator and myself. I drew otters, read about them and even dreamt about them. As soon as I was able, I would disappear down to the river Ouse on the edge of our town to search for the shape of the world that I had read about.

I never found any sign of otters. They weren't just scarce where I lived; I heard that in many areas there truly were none left. This was the late 1970s. Many of the rivers in the highly cultivated English landscape were devoid of suitable habitat. The post-war land-management policy had been to increase food production by any means. The rivers were cleared, straightened and canalised, and then were made toxic with run-off and residues from agriculture. DDT, dieldrin and aldrin from sheep dip and seed-dressing were liberally spread about; lead from petrol seeped into the water, along with who knows how many other chemical pollutants. These magnified as they passed through each successive layer of the food chain, concentrating in the tissues of the top predators. Animals like the otter and the peregrine falcon were at the end of the line, and the effects of these toxins were devastating. Peregrine falcons' eggs became too fragile to hatch. Otters, living in an aquatic ecosystem that collected every single poison we humans had invented, went blind, or failed to produce young, or both. With no reproduction possible, slowly, quietly they began to pass out of existence.

Otter-hunters, who had the most accurate idea of otter numbers, were the first to notice the sudden decline. It was when I discovered that the otter was not protected that I joined the campaign for its preservation. I became a junior member of a new foundation called the Otter Trust, a pioneering reserve run by otter experts Philip and Jeanne Wayre. Finally, in 1978, a special newsletter from the Trust announced that the otter was now legally protected in England and Wales. There was more good news: 'Kate', the European otter, had given birth to three cubs and the Wayres were continuing with a careful breeding programme, before reintroducing captive otters into the wild. Otters from Scotland, where they were still relatively common, were brought south to boost the population. During his work with the otters, Philip was able to observe them at close range, even sometimes swimming with them underwater. He described their habits in his book *The River People*, which I bought as soon as it appeared. Now, with the help of the Wayres, at last I could see a live otter for myself.

I persuaded my mother to take us on holiday to visit the Otter

Trust. We set off, my long-suffering mother, my teenage brother and me, on a trip that was to cement the longest love affair of my life. When we arrived, I shot out of the car like a bullet and barged through crowds of visitors to find a place at the wire-mesh netting. Fingers gripping the fence, I watched the otters play and dive sinuously under the water. They gambolled around inside generous pens and, as I watched them, my heart swelled in my chest. I ran around each enclosure, butterflies in my stomach, noticing that each otter had a name, or a home originating somewhere far away. Philip Wayre was bringing back the wild otter and, as far as I was concerned, he was a hero.

Back home again, I collected scraps and cuttings from the papers about the decline of the otter. I developed my own campaign publication called 'Otter News'. It had strident editorials, helpful, otter-related information and a readership that encompassed my friends, my family and much of my street. I still have four surviving issues. They are an archive of material, and capture a moment in history when the otter had declined to critical numbers. In their own small way the scraps I gathered for Otter News track a moment when science, the conservation movement and the public mobilised themselves for a common cause. In those early days the conservation methods were haphazard and there were disagreements, but a world without otters was unimaginable, and many people agreed on that.

Now the otter has become an icon of nature conservation. In our minds it is cuddly and playful, but in reality it lives the harsh life of a top predator. It represents the worst and the best of what we humans can achieve in our relationship with the natural world. It is a barometer for the health of our rivers, and like a spirit level, its reappearance indicates that we are finding our balance with the wild, and we should always remember to keep a weather eye upon it.

Other than the snatched sighting in Scotland, I'd never seen a wild otter. That brief sighting sparked something in the core of me. Years of hoping and imagining, miles of river and road, and finally I found one, out of my window. Now I'd had a glimpse, I wanted to get closer to this creature which had filled my imagination for so many years.

A plan formed in my mind. I would explore the places in this land that hid my grail. I would spend a whole year or longer, if that's what it took, wading through marshes, hiding between mossy rocks, paddling down rivers and swimming in sea lochs; recording my journey through the seasons as I searched for wild otters. I would embark on a water-level odyssey around our small archipelago, following another map of Britain: the undulating mesh of blue veins that is Otter Country.

True North

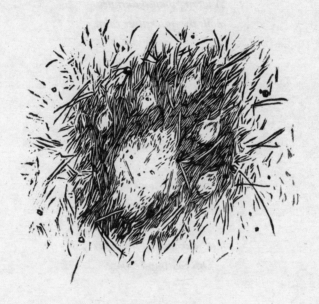

A wrong-shaped darkness
moils the quicksilver calm,

tumbles over itself,
sculls in and out of vision,

then with a slick tail-whip
is gone, leaving bubbles

(you imagine), widening rings
and a new marking

onto your map of this place:
Otter-Glimpse-Shore.

Chris Waters, 'Once Given'

It's August and the air around my home in Devon is stuffy and humid. The June trip to Scotland seems a long time ago and the days are beginning to draw inward. Where I live, otters are so elusive that their activities are entirely invisible. If I want to get close to one, my instincts tell me that I'll have to travel back north. The sea lochs are the only place I know where daylight can reveal the gleam of water on fur, the flash of strong jaws and sharp teeth on prey, and the silky movement of the otter as it swims.

For people who are interested in wild things, otters have a unique and complex magnetism. They are wild and not wild. They seem to have an alluring softness and yet are one of our most effective top predators. Rarely seen in daytime, they are sublimely secretive. They can appear and disappear at will. The creature inside us is provoked by the otter's dual nature and drawn to its playful lightness. We admire its amphibious skills, its resilience and its ferocity.

If I'm going to see another one, I need to improve my tracking skills. I also need to shed the ideas books have given me. Most of the time we frame these animals through the cloudy lens of our imagination. Outside the specialised, scientific world, much of what we think about otters is mediated by television and by books like *Tarka* and Gavin Maxwell's *Ring of Bright Water*. The soothing image of these animals enchanted millions, and in the 1960s keeping pet otters made Maxwell famous. His lyrical account of life with otters in the wild emptiness of the Highlands was highly seductive, but how much did it actually tell us about otters? The enraptured descriptions reveal nothing of the conflicts that emptied the land of its people and little of the reality of keeping wild animals in captivity.

I plan to find the true 'Camusfeàrna', where Gavin Maxwell lived. This place is as remote as any in Britain, and as well as red deer, golden eagles, wildcats and seals, I'm sure there will be otters. Camusfeàrna – in reality called Sandaig – is on the west coast of Scotland, on the Sound of Sleat, a fast-flowing channel between the rocky edges of the mainland and the mountainous Isle of Skye.

August is the peak of the midge season in the Highlands, but neither Maxwell's book nor the romantic film that it inspired mentions this biting scourge. I'm warned that it will be far worse than in June, and head off prepared for combat, equipped with midge hood and long trousers. Parked in a lay-by on Rannoch Moor I hear a menacing hum coming from all around. As soon as I open my window I realise that it will be prudent to close it quickly and stay inside. The people who toiled here before lived with many things that were out to kill them, and for the midges they used a strong-smelling plant – bog myrtle – to protect themselves. I'm not going to risk being savaged for the time it would take to gather these aromatic leaves, so I continue my drive north-west, via miles of moor and lapwings, to the rolling Atlantic of the west coast and the Isle of Skye.

My first night in the mountains is a baptism of midges. The swarms are too extreme to sit still looking for otters, so I give in and seal myself inside the tent, daubing my bites with antihistamine. It is far too early to sleep, so I settle down with some oatcakes and my new edition of Maxwell's *Ring*. Early in the morning, I take my battle scars to the coast at Glenelg, and instead of choosing the road bridge across to Skye, which might have saved me from more midges, I can't resist riding on the tiny turntable ferry just a few miles from Maxwell's home. I've heard there's an otter haven beside this crossing to Skye and a hide where you can watch wild otters fishing in the Sound. Travelling alone, I find myself babbling through the midges to anyone who will listen, in this case the ferryman. I suspect he is plastered in some secret highland deterrent, as he seems immune to the ferocious insects. Later I discover that one of the best protections is Avon moisturiser. It is not sold as repellent, but midges can't stand it. I try standing inside the ferryman's midge-free zone but it has no effect and the swarms gather hungrily around me.

The ferryman notices my RSPB and Wildlife Trust window stickers, and we chat about birds. He tells me about a sand martin colony nearby. We discuss changing ocean currents and their effect on sand eels, and the gradual disappearance of puffins and other seabirds. I tell him about my otter search. He points south, to the bay on the mainland where Maxwell made his home. I want to know the best way to get there. The ferryman, who seems to be knowledgeable about everything, and a serious ferry enthusiast to boot, explains the route south to Sandaig. The way soon becomes visible from the water, and I try to memorise the intricate directions as the ferry swirls this way and that over the green-glassy gulfs of the water of the Sound.

A regular ferry was already well established here when Boswell and Dr Johnson toured the Hebrides in 1773. It is thought that they stayed in the Ferry House, a pretty, whitewashed house which still nestles a few yards up the road from the quay. This ten-minute crossing must be the shortest and most spectacular in Britain. As we dock, my attention is drawn upwards to the steep slopes of Skye, where everything blends into a glimmering camouflage of rock, bracken and heather, concealing all animals from red deer to eagles. Heavy cloud and shafts of sunlight dapple the tweedy greens, purples and browns of the slopes. On the mainland shore only a few tiny croft houses dot the landscape; on the island there are even fewer. It amazes me how anybody could make a living in this isolated spot, but the tiny roll-on-roll-off ferry and the busy Skye bridge prove that many do. The ferry has been taken over by locals to preserve the important crossing point, which is the narrowest one between Skye and the mainland.

Seals bob like vertical bottles, playing around in the choppy water that rushes through the narrow strait. I scan the surface and rocks for signs of otter. For a moment I think I see the slip of liquorice that could be the curve of an otter's dive – slimmer and shyer than the bold profile of a seal – but in a blink it has gone.

Otters, seals and birds all play tricks with water; the marine animals' smooth disappearances invite the imagination and conjure beings to fall in love with – sprites, selkies, mermen, sirens, kelpies –
stories made to seduce humans into this mysterious other world.

'We see otters from here every day,' the ferryman tells me. 'They're
used to us. They rely on the shift of the tides rather than the night-
time for hunting.'

I plan to pitch my tent close to the water and begin my search at
dawn. 'You should talk to Jimmy,' he calls as I leave the ferry. 'He lives
in that house back there.' My heart leaps. Could this be Jimmy Watt,
Maxwell's first otter-keeper? The ferryman confirms it.

I make a calculation in my head about Jimmy's age. His photograph adorns the front cover of my copy of *Ring*. It shows a young
man, but it was taken about fifty years ago. I flip through my new
paperback, looking for more photos. Would I recognise Jimmy now?
There are many photos of him in this new edition, a trilogy with the
two books that Maxwell wrote to follow it, *The Rocks Remain* and
Raven Seek Thy Brother. In the pictures, Jimmy is always doing something; on the cover he is in a small rowing boat, with wild mountains
and sea all around him. He has a floppy fringe, rolled-up jeans and
looks about fourteen or fifteen. Edal, the second pet otter to live at
Camusfeàrna, is leaning on the side of the boat, wearing a harness
made specially for her.

I have become more and more curious about life with Maxwell
during my frequent rereading of passages in the book in the preceding
week. I know that he was a brilliant writer and that he loved otters,

21

but what can it have been like living in his shadow? He certainly didn't treat all animals kindly; foxes, basking sharks, orca, red deer and grouse all suffered from his energetic hunting habit. Although he claimed to love living creatures, not all were treated with the same affection that he had for otters. I'm not sure about what he was like with people, but I have heard he could be difficult. Now I find that one of the heroes of the story, and one of the people consistently close to Maxwell, is right under my nose.

Will I go? It's very tempting. He may be happy to have a conversation; he may not. At least he might sign my copy of the book. How can I know unless I knock? Visiting Jimmy unannounced could be an unwelcome intrusion, so I decide to go to the otter haven close to the ferry crossing at Kylerhea on Skye, and to sleep nearby, giving myself time to think about what to do. I find a place to pitch my tent, on a smooth, grassy outcrop just across the Sound of Sleat from the legendary Camusfeàrna.

The midges are on good form; the wind has dropped to produce perfect feeding conditions again, and there is nothing to protect me from these minute, remorseless blood-suckers, but I too must get my fix.

Thousands of people come to the otter hide at Kylerhea each year, just to catch a glimpse of an otter swimming in the Sound. Inside there is plenty of seating, with stealthily placed windows from which to watch the water. When I arrive there is a family who repeatedly shriek '*Otter!*' at fever pitch, each time a seal pops up. I hope they won't stay too long. Otters have excellent hearing and are particularly sensitive to any sort of rumpus. Seals on the other hand are not always as fazed by noise, in fact they can be attracted to it. Seals swim underwater for longer than otters, and can appear alluringly large as their whiskered faces and inscrutable eyes rise out of the water. They seem as curious about people as we are about them. It's tempting to think an otter might share this characteristic, but usually otters are not interested. You could think of seals as the dogs of the sea and otters as the cats. A dog is usually confident and sociable, whereas cats can be diffident or wary. Seals will stay immobile for quite a while, heads above the water, goggling at any people within close range. They will sink when they have had enough, or porpoise with a copious splash. Seals have tiny ears that are like openings, quite close to their eyes. Otters have more prominent ears, set further back on a flatter head, and they are far lighter and more delicate than the buxom bulk of a seal. From a distance it is possible to mistake the dive, but otters have a purposeful, busy aspect to their fishing which is unlike any seal in the water. Seals watch; otters fish, dive and spar, and the thin tip of the tail often appears beside them. They generally seem to have a more nervous, feline sensitivity about them, and will not often wait around or stare back if they know they are being watched.

After the family leave, I sit quietly with a woman who has never seen an otter. We watch and wait, sharing binoculars. Eventually one appears in the rising ripples of the tide, foraging in the clear water around the lighthouse. We sit and hold our breath for a moment, watching the otter's lithe form porpoise again and again under the surface of the waves. It pops up with something in its jaws and comes ashore to crunch its prey, then it dives in again for more fishing. It works the coastline meticulously, searching every pool and trench, and the water is so transparent that we can even see it underwater as it twirls and blends into the fronds of kelp. One more time it surfaces with a fish, this time holding it between its front paws to eat. Now the webs and claws are clearly apparent as they clutch the fish in a snug grip. When the last morsel is swallowed, the otter dives again. We catch a final glimpse as it disappears around the lighthouse and then it is gone.

Otters give so little of themselves it's hard to know them. We may see a cryptic bump in the waves, a whiskery snout; then nothing. They can stay underwater for a few short minutes, usually two or three at most. But otters are hard to measure. In rivers they can sink imperceptibly and resurface unseen in another spot. While underwater they can swim extremely fast, and this also explains the mysterious disappearances. Valves in their nostrils and ears close to keep them watertight, and a trail of bubbles can sometimes be seen rising to the surface. This can be the only sign the otter is there. The air bubbles that appear are not all exhaled from the otter's lungs, but are squeezed sporadically from the mouth and fur as it pushes with all four webbed feet down into the pressure of the water.

Sometimes the otter will porpoise unexpectedly. This looks like play, but it could have a more practical purpose, to increase the otter's speed or to help it to dive to a greater depth while fetching prey from the seabed. It might be that when the otter has been foraging underwater for a while, the trapped air inside its coat may have been forced out, and the outer fur becomes waterlogged. The otter needs the air for insulation, so to replace it he might occasionally shoot out of the water in mid-hunt, in a deliberate arc, to aerate his fur whilst on the move. As with many aspects of otter behaviour, experts don't entirely agree. It is possible that all the theories are right, and as usual, the otter's behaviour defies a single, fixed explanation. One thing is certain: an otter's beautiful buoyancy and connection with its element imbue it with the appearance of lightness and joy.

The otter belongs to water. Its name comes from the Old English *otor*, and from the same root word as water. Both the Latin name – *Lutra lutra* – and the English name echo the Greek for water, *hudor*, and the Latin for wave, *unda*, linking the animal and its onomatopoeic name to the bubble and flow of its watery world.

Inside the hide on the Isle of Skye, we wait for the wild otter to return. As I scan the shoreline with my binoculars, every so often my heart misses a beat. I'm caught out by floating rafts of weed, the curve of a cormorant, otter-shaped rocks, wavelets, mallard, oystercatcher. Somewhere he will be sunk in the water, perhaps even watching us, the water lapping around his nostrils.

He never reappears. Later, I go down to the sea and swim. In the freezing cold, the dark water is full of swirling currents and jungles of weed. Every molecule of my blood thrills and my breath is squeezed out in one shout. I think of the poet Alice Oswald's watery description of what could be beneath me: 'In each eel a finger-width of sea'. The thought of the unfathomable depth and the unseen fins, pincers and jaws leaves me tingling.

Swimming in the sea is like sliding into an imaginary world, but it also gives an otter's-eye view, where I'm on more equal terms with water. As the green cold seeps into me, I'm no longer towering above and separate but embodied within it. I come out feeling as if I have shed something. I am lighter, and more myself than before.

Sometimes people who have been in the West Highlands report that they have glimpsed a 'sea otter'. If they have been lucky, and they have wiped their binocular lens properly, and it was not a seal or a water bird, they have actually seen the Eurasian otter. The thirteen species of otter that we know about live on almost every continent of the planet except Australia and Antarctica. We only have one species in Britain, and this one is found in many parts of Europe, North Africa and across Asia, so although it is scarce in many of these areas, it actually has a massive range.

Lutra lutra was classified over 250 years ago by the great Swedish zoologist Linnaeus in his *Systema Naturae*, and in Britain is the largest of the Mustelidae, or weasel family. This group includes the weasel, stoat, polecat, pine marten, badger and mink. The mustelids are generally short-legged, often extremely strong for their size, and can be highly supple. The family is characterised by anal glands which produce a musky scent both for marking territory and for sexual signalling. Scientists believe that the otter's mammal ancestors evolved about forty million years ago. They diverged into the pinnipeds and the mustelids about twenty-three million years ago. The pinnipeds were mammals with flippers, like seals, which could live in the open ocean, while the mustelids lived on land and, in the case of the otter, on rivers and shores as well. Millions of years before we appeared, the otter already had webbed feet and was semi-aquatic.

The otter's generous size and thick fur allowed it to forage in water at low temperatures. This broadened its spectrum of prey immensely, and it could feed on all sorts of aquatic species, increasing its chances of survival. After seals and cetaceans (whales, dolphins and porpoises), the otter is one of Britain's largest predators, an adult reaching up to 125 centimetres long and weighing up to ten kilos.

We are told that the otter is territorial, and it is true that it needs a large amount of aquatic territory to feed itself, which means it must be solitary. The male otter has to defend the females in its range from other males, but his range might fluctuate. A pair or a group would have to travel much further to find enough food, so this is why each otter must live in a solitary way. Having said this, the otter can be quite sociable with other otters that it knows. It could be that otters live in diffuse 'tribes', only fighting with outsiders who might compete for mates, dominance or food.

As they evolved, all of the mustelids developed subtly divergent feeding, which minimised territorial competition. These different habits are fascinating in themselves; today's pine marten and polecat, almost as elusive as the otter, are the arboreal members of the family, feeding on a range of foods found in trees, from small mammals to berries, birds and eggs. The badger is terrestrial, burrowing into the earth and foraging on the bulbs, roots, worms, slugs, invertebrates

27

and small mammals it finds there. The weasel feeds on voles, mice and insects. It is not uncommon to catch sight of a stoat trotting across a road, carrying a large vole in its jaws, but they can also be seen hunting rabbits. Considering that rabbits can be more than ten times the size of a stoat, the stealth, strength and stamina required to hunt this prey are impressive and often involve elaborate acrobatics. The stoat has a special weapon: its canine teeth are perfectly evolved to give a precision bite that will dislocate the vertebrae of the rabbit's neck and kill it instantly. Weasels are equally well equipped, although the smallest in the family. The phrase 'to weasel' must come from humans having observed this mustelid's enormous energy and cleverness. The weasel seems never to be still, darting about like lightning and going to prodigious lengths to seek out and capture prey.

The mink, a larger and more furry version of the stoat, but still only half the size of the otter, also lives aquatically. It too is a ferocious predator, extremely stealthy and quite versatile; once it has exhausted one source of prey it will move quickly on to the next. Conflict arose in the delicate balance of the ecosystem when this non-native mustelid from North America was introduced to Britain. From the 1920s mink were farmed for their extraordinarily dense and luxurious fur. By the 1950s there were at least 400 fur farms across Britain and, following sporadic escapes and releases, a feral breeding population quickly established itself in the wild. Many escaped from where they were bred, and increasing numbers were released into the wild as the fur industry collapsed. They have caused trouble for every species with which they came into contact and in Britain mink have consequently had a very bad press.

Otter and mink compete for prey and may battle fiercely if they come into contact. The otter is twice the length and ten times the weight of a mink, so it must be a daunting opponent. But the mink can be a threat to the otter; where food is short there is likely to be violent fighting, and infected bites from mink and even dislodged mink teeth have been found in otter corpses.

The poor water vole, better known as 'Ratty' in Kenneth Grahame's *The Wind in the Willows*, has been the worst victim in the affair. This shy rodent unfortunately resembles the muskrat, the American mink's most common prey. The mink is a voracious hunter, and exactly the right size to slither into the water vole's small riverside burrow, easily devouring anybody inside. In some areas the water vole has completely disappeared, and in others, populations have been reduced by as much as 90 per cent. To remedy this, affected parts of the country have 'controlled' mink by extensive trapping.

Perhaps the most impressive mustelid is the wolverine, which is no longer found in Britain, but deserves a mention. Its stocky, bear-like body has super-insulated snow-shoe feet, a thick, oily and entirely frost-resistant coat, and it can travel similar distances to the grizzly bear. Fossil remains of wolverines, dated to about 83,000 years ago, were found in caves in the Yorkshire Dales; these were discovered alongside remains of bison and reindeer, which they may have been feeding upon. The wolverine must have found its way to Britain during the last glacial period. Its special feature is a pair of strong back molars rotated at ninety degrees, perfectly designed for tearing bones and meat that have been frozen solid. Arctic travellers and Polar explorers once used wolverine fur for clothing, as it was by far the warmest and most protective of animal pelts. With all these indestructible features this animal is the mustelid tough guy. As the ice sheets retreated in Britain, the wolverine followed. The only evidence it left behind was its bones, remains of the colder, harsher environment that was once here.

All mustelids are outlandishly strong, and our Eurasian otter is no exception. As anyone who has handled one and been bitten will tell you, it is well served with bone-crushing teeth. These have adapted to deal with the tough shellfish it finds by coasts and estuaries. But despite its coastal foraging the Eurasian otter does not actually live in the sea. Our otter could not survive for long in a purely salty environment; it needs a constant supply of fresh water to drink, and to rinse and clean its fur. Without this it would quickly lose its waterproofing, develop hypothermia and die.

The true sea otter, *Enhydra lutris*, is exclusively marine and has adaptations that allow it to live permanently in the sea. With the densest double-layered fur of any mammal, at up to 165,000 hairs per square centimetre, an adult male can average a record-breaking eight hundred million hairs over its entire body. Our British otter has less than half that number. The sea otter has a harsher environment to deal with, as it is native to northern and eastern Pacific coasts. It's larger, chubbier and even more buoyant than our otter. The bushy, gold-tinted pelt is the longest of all otter species, and beneath its layer of silky 'guard hairs' it has an ultra-dense layer of insulation. This is formed by matted bundles of woolly under-hair, creating a super-thermal and almost impenetrable body protector. Because of this, the sea otter is able to live entirely out at sea, feeding purely on the shellfish that thrive in deep underwater kelp forests. Sea otters are an important part of the marine ecosystem, maintaining the kelp by diving down to feed on the sea urchins that graze it. Scientists studying animals' effects on climate change have discovered that by protecting the fragile kelp forests, sea otters actually help to reduce carbon emissions.

Sea otters may need to dive down a great distance to get to their prey and they have an enlarged ribcage to deal with the pressure at these depths. Their teeth are specialised as well, with two sets of incisors, as opposed to the three pairs that fish-eating otters have. The other set of incisors has been replaced by extra molars to crush and grind the hard shells of the crunchy prey available to them.

Bobbing in deceptively cute family herds or 'rafts', each sea otter is like a small grizzly bear in the waves. Their jaws are monumentally strong and sometimes ferocious fights take place. To sleep afloat they wrap themselves in a scroll of kelp to keep from drifting away. The kelp conceals them from predating orca and great white sharks who would not hesitate to lunch on one of these floating morsels; unfortunately the sea otter's teeth are no match for these gargantuan predators.

The only sea otter we see in Europe looks out from postcards and wildlife calendars, floating happily on its back, often clutching a clam to its creamy breast. The idyllic picture is doubly deceptive. The happily floating demeanour, with flipper-like paws held aloft, is in fact not laziness but a trick to keep warm in the freezing Pacific Ocean. What is more, because of its rich, desirable fur and trusting disposition, it has suffered merciless persecution from humans. It evolved to live in the sea to avoid us, but by developing the super-thick fur necessary for this environment it only seems to have attracted more attention. In 1800 European settlers arrived on the Pacific North West coast of America to hunt the sea otters for their skins. Stories are told of how the otters innocently nuzzled hunters who were about to kill them. Hundreds of thousands were killed, along with many First Nation people. Without the reverence or understanding of the local people, greedy European hunters blithely set about wiping out the species. Within a hundred years the animals were so scarce that a single pelt was worth as much as a thousand dollars. The Californian sea otter's fur was known for a time in the early nineteenth century as 'soft gold', and very soon the species was hanging on to survival by a thread.

In 1915 a tiny cluster of surviving sea otters was discovered just off the rugged Californian coast at Big Sur, a fact kept quiet by the authorities to let the population recover. Although they are now a protected species, oil spills, along with other toxic pollution, still exterminate them by the thousands. The population took another dip more recently, at the end of the twentieth century, as whaling and over-fishing upset the entire ecosystem. Orca attacks are common, and as if this were not enough, the poor creature suffers from a weak heart, caused by its habit of balancing rocks on its chest and bashing its favourite clam meal upon them until the shell breaks open. Such frequent blows to the heart are not without consequence.

Female sea otters are even worse off. The male otter mates very aggressively, biting the female's nose during copulation; she will often be permanently scarred by this, or even killed. In an even more brutal habit, the males kidnap the young from their mothers and expect her to hand over her catch of shellfish before the stolen pup is returned.

Sea otters are born out in the waves. Their silky fur has a quality so dense and yet gossamer-fine that it is hard to tell where the softness ends and the air begins. The pup is carried close to its mother's body, kept safe and warm, and licked until it is so fluffy that its fur contains enough air to float like a cork. Pups are carefully guarded until old enough to cope with the temperatures of the ocean's cold cradle, and until then the mother lies on her back in the waves and holds her young one on her front, with her forepaws clamping it safe from any danger. The bond continues in this way for as long as the mother otter is able to hold on to her young.

The Californian sea otter has attracted such attention and sympathy in recent years that fourteen million people tuned in to watch a YouTube video of a mother otter, rescued from an oil spill, 'holding hands' with her pup in the Monterey Bay aquarium. In the period following this, the Californian State Tax voluntary contributions for the conservation of sea otters reached one million dollars. The same year, in the marina in Monterey Bay, yacht owners and fishermen watched astounded when a lone female sea otter gave birth to a pup on the dockside.

Sea otters are normally shy, but not this mother. Clutching the pup protectively on her belly, the mother otter had decided to use the wooden boardwalk as a nursery. Stray coils of rope were substituted for kelp cradles as the baby otter grew. Soon, a strange new sound echoed around the keels of the million-dollar yachts. It was the mother otter, tap-tapping her shellfish against the boats. The resourceful otter was using the undersides of these expensive craft as nutcrackers. They were perfect for teaching the new pup to break open its abalone and clam dinners. She risked her life as she chipped away those thousands of dollars' worth of paint, but as the pup learnt to feed, something more profound was happening. This mother otter was chipping away at our estrangement from wild animals. The fragile young family was made welcome, irrespective of the damage to the boats.

So why did the mother otter give birth amongst humans, whom she would normally regard as deeply dangerous? The harsh environment in which sea otters survive may provide the answer. These animals live on a knife-edge. A bundle of fluff the same weight as a bag of sugar does not seem suited to living in the thirty-foot storm waves and chill temperatures of the Pacific Ocean. If it were left alone in the water, the pup would quickly freeze to death. Mother sea otters carry their pups permanently on their bellies to avoid this, but at some point they need to leave their babies and dive underwater to feed. Was food scarce for this mother sea otter? Or was it the threat from the otter's natural predators, shark and orca? Whatever the answer, meeting the gaze of this shy wild animal touched many people and gave us some more understanding of her precarious lifestyle. And as each boat-owner motored carefully past, they felt the piercing and watchful gaze of the mother otter as she protected her child.

At the beginning of *Ring of Bright Water* Gavin Maxwell describes the first time he is able to observe his otter Mijbil swim underwater in a home-made glass tank at Camusfeàrna: 'His speed was bewildering, his grace breath-taking; he was boneless, mercurial, sinuous, wonderful. I thought of a trapeze artist, of a ballet dancer, of a bird or an aircraft in acrobatics, but in all these I was comparing him to lesser grandeurs; he was an otter in his own element, and he was the most beautiful thing in nature that I had ever seen.'

Mij was not Maxwell's first otter. It had taken some effort and a few false starts to find one. The first otter was given to him in 1956 by the British explorer Wilfred Thesiger while they were travelling together in southern Iraq. Initial searches around the Iraqi marshes came up with a very young female otter who was named 'Chahala', after the river where she was born. Maxwell describes choosing the name after listening to her crying at night; the cry sounded just like the name of her river. But disaster struck, and the little otter died after Maxwell unwittingly fed her fish that had been poisoned by local fishermen. *Digitalis* had been used to drug the catch; the tiny doses were harmless to people but lethal to fish and, as it turned out, small otters.

After this first failed attempt Maxwell redoubled his efforts to find an otter to bring home. He describes the moment he was given a new cub still squirming inside a bag: 'With the opening of that sack began a … thraldom to otters, an otter fixation, that I have since found to be shared by most other people who have ever owned one.'

Otters may be enthralling, but they make terrible travelling companions. During the highly precarious journey home, there were indications of what was to come for both otter and owner. It was a serious problem to keep Mij contained, and neither Mij, Gavin Maxwell, nor many of the airline passengers touched down unscathed. On the aeroplane Mij was so distressed inside his box that the kind but misguided stewardess suggested he travel on Maxwell's lap. She evidently knew nothing about otters and laps: 'He dodged my fumbling hands with an eel-like wriggle and disappeared at high speed down the fuselage of the aircraft.'

Any wild animal released into an aircraft cabin would cause some degree of confusion, but an otter more than most: 'I could follow his progress among the passengers by a wave of disturbance amongst them not unlike that caused by the passage of a stoat through a hen run. There were squawks and shrieks and a flapping of travelling coats, and half-way down the fuselage a woman stood up on her seat screaming: "A rat, a rat!"'

Once they were settled in Scotland, their new life together was compellingly described, and Maxwell became fonder of his otter than of almost any human being, even after it had rampaged through his London flat, sunk its bone-crushing teeth through his hand, daintily pierced each of his earlobes and caused no end of domestic chaos.

After only a year Mijbil was tragically killed, but undeterred, Maxwell pined for a new otter. He found a female, owned by a doctor who worked most of the year in Africa. Named Edal, this otter had an unusually abundant soft pelt. The photographs of her in *Ring* show that she was probably not the same species as Mij. Her nose and facial markings suggest that she might have been a Cape clawless otter, or *Aonyx capensis*. This species inhabits the Niger Delta region of West Africa, where Edal was found. She was bought as a tiny cub from local fishermen by the Benin River, and her British owners brought her home with them, before passing her into the safe keeping of Maxwell.

Though far from her native country, she appeared at first to have adapted well to her new home. Maxwell describes the new otter nuzzling at the nape of his neck: 'that well-remembered, poignant touch of hard whiskers and soft face-fur ... Edal became mine, and there was once more an otter at Camusfeàrna'. There is touching film footage of Maxwell playing affectionately with Edal, and the close bond of trust between them is clear, but from an early stage Edal needed a lot of attention. Maxwell travelled frequently and used his Highland home as a retreat, so he could not look after his animal alone. Two boys, Jimmy Watt and later Terry Nutkins, were brought in to work as the otter's keepers. Maxwell's friend Richard Frere described the new recruits as 'lusty and jubilant young men, of Viking stature and Nordic good looks ... at an age when everything at Camusfeàrna was seen in boyhood's visionary gleam as a rough playground spiced with adventure and endeavour'. It sounds like an idyll, but Nutkins later revealed that he felt that Maxwell lavished all his affection on the animals, creating a bond that seemed stronger and easier than any of his relationships with people.

37

As Edal grew to maturity it turned out that otters do not make good pets. She had to be fenced in, and Teko, a new male otter chosen as a mate for her, had to be kept separately. She did not appreciate the new company and had become increasingly volatile. If the otters met they fought, and they even attacked the boys and Maxwell's friends. As they had each savaged their keepers, great care had to be taken around them. Richard Frere described his first impression of the otters as slimy and sinister. 'With tempers so unpredictable, and jaws so strong – I saw them crunch a sizeable fish in half with one bite – it was no surprise that even those who knew them approached with the utmost caution,' Frere wrote. Many friends were badly bitten; Nutkins himself lost two fingers, caught a gangrene infection and was hospitalised due to the savage and capricious side of Edal's nature. Frere was surprised that Nutkins could even bring himself to care for 'the creature whose ferret-like ferocity had robbed him of his fingers'. Nevertheless, images of the adorable otters abound in *Ring*. Edal is captured lying voluptuously in front of a nude odalisque painting; Mij is seen juggling with marbles. Both otters pose in armchairs, curled or sprawled asleep in a bed, and Teko is portrayed trotting at the end of a lead, or cuddled in his master's arms. Maxwell must have believed that in showing his otters like this he would bring the species into the hearts of the reading public; he may have been right, but harnessed, tamed and cosseted, something of their wild essence was lost.

After my swim, I walk to warm up. Every part of me feels clean and alive, as if the seawater has soaked up some of my restlessness. The world is crackling with possibilities. Somewhere out there, even further north, is the Isle of Lewis. To the west are the dramatic peaks of Skye. It is a long and arduous journey to get anywhere up here, but even on this far edge, a little way round the coast, there are signs of human activity. Heaps of detritus and jetsam have travelled here, washed up in tangles of plastic. Colourful netting and rope, tar, cans, boxes, stray fishing buoys, odd trainers, toothbrushes, plastic gloves, polystyrene, torn shopping bags, broken barrels and worn-out bits of boat lie everywhere. I try and fail to count the extraordinary number of empty bottles that seem to have drifted together on the ocean currents and arrived here in jumbled rafts. On the sand at the tide line there seems to be a plastic bottle top for every colour and shade of the rainbow. I spend some time collecting tops, and find more colours than I can name. Beside these items, splashed on the rocks like a riposte of colourful graffiti, are sea lichens, crab shells, layers and layers of barnacles, limpets, anemones and piles of living seaweed and uprooted kelp.

Back in the van, the news on my radio reports floods. Winter weather in summer, snow and hail in drifts, rivers that burst their banks, storm drains failing, a cathedral threatened. The talk is of apocalypse, but it seems to me that it is just water balancing itself out. The moon pulls this way and that, and water rights itself. We build outflow pipes, ditches, culverts and canals. We tarmac and pave our roads and driveways so that water cannot drain in the proper way. Yet in some parts of the world, when heavy rain comes, people perceive water as simply animating the landscape; as long as they are not washed away or flooded out, people play, tell stories and celebrate. We dream of water; sometimes it is the stuff of our nightmares, when nature has its say, but it is also an element of imagination, and the wild otter is like a living current playing within it.

Maxwell witnesses hints of this as he watches Mijbil when he first discovers the bath:

> ... he went wild with joy in the water, plunging and rolling in it, shooting up and down the length of the bath underwater, and making enough slosh and splash for a hippo. This was, I was to learn, a characteristic of otters; every drop of water must be, so to speak, extended and spread about the place; a bowl must at once be overturned, or, if it will not overturn, be sat in and sploshed until it overflows. Water must be kept on the move and made to do things; when static it is as wasted and provoking as a buried talent.

The wild otter lives in a cold, dangerous waterscape that continuously pours away and remakes itself. Its watery labyrinth is an unpredictable world. The tides, pools and riverbeds bring as many inscrutable risks as they do possibilities. Electric pulses in the river beckon, the buffetings of water and weather are always a challenge and perhaps only sometimes a pleasure.

The wild otter I saw would no doubt be out of the water and making tracks to its own musky holt, to curl belly upward, in a home of roots, peat and rocks. I imagine him enfolded in his fur, dreaming of water; a tight sleep-knot, enjoying the deep sleep of one who exists totally in the moment. An otter may spend more than twelve hours asleep inside its holt or nestled on a couch of grass each day. His coat is double-layered so that he never feels wet against his skin; he is felted with a woolly undercoat which insulates and wraps him, and covered on the outside by a darker duvet of protective fur, guarded from even the lowest temperatures. In my own bed, encased in four walls, double-glazed and muffled from the outside, I still cannot hope to experience the solid rest of that wild otter. Everything conspires to keep me awake: my dreams are fitful, weighed down and polluted with worries; the flotsam of relationships, memories and the clutter of possessions.

The otter has none of this. Outside breeding times, it doesn't even have a home. Living a nomadic life, a dog otter may have many different places to sleep, spread over up to twelve miles of territory. The female may have fewer places; her holt is usually hidden along a riverbank amongst thick brambles or in the dark amongst the roots of a tree. A den might be a vacated fox burrow or badger sett; in Scotland an otter will burrow into the soft peat or enlarge another animal's vacant home. When I say vacant, this may well not be voluntary on the part of the original owner; to another mammal the otter could seem larger and bolder than we might think. A fully grown dog otter can almost match a badger in terms of size and weight, and its jaws are strong enough to break a large crab or snap the spine of a fully grown salmon.

41

Sometimes the otter will sleep on a couch of grass or flattened sea-weed, right out in the open. I have noticed these sleeping places on land. Often, near the otter's holt or couch, there will be a rolling place where it has spent time meticulously drying its fur before sleep. The otter will roll over and over like a cat in the sun until every hair is smooth. There may be spraint, the otter's droppings, nearby, and these signs can sometimes form great mounds. They are a dead give-away that an otter has been on regular patrol.

The wild otter must keep moving to survive. It has very little body fat and must work every muscle to the peak of condition. When sub-merged in water it seems to be weightless, it can lure the eye then disappear, leaving you unsure what it was or whether it was there in the first place. It can swim right at the heart of an approaching wave with no fear, but has no concept of the danger of fishing nets and oil slicks.

The otter no longer suffers the same persecution from humans as it did in the past; hunting has ceased, and it is illegal to keep one as a pet. The otter has no natural predators and has become an object of fascination, perhaps above all other mammals in Britain. Beautiful, intelligent, adaptable, it is also as savage as it is alluring. It's no surprise that it has been coveted, nor that human admiration for the otter's skills and beauty have been translated into the some of the most cap-tivating film, prose and poetry ever written about a wild creature. This is a double-edged sword. Otters do not want to be known. For wild animals such as this, close contact with us often spells misunderstand-ing, or at worst, disaster.

Two of Gavin Maxwell's otters died violent deaths. In the late 1950s, wild otters were generally seen as vermin, and their pelts would have been worth a good deal of money. Legend has it that Mijbil was murdered by a road mender who mistook him for a wild otter while Maxwell was away from home. The appearance of an otter roaming slowly along a roadside ditch must have been like finding a pot of gold, and too good an opportunity to miss. In *Ring* Maxwell writes about the incident. He found out that Mij had been bludgeoned with a pickaxe by a man in the nearby village. The culprit's name was changed for the purposes of the book to 'Big Angus', but the true story was not a secret and caused shock waves locally. 'I arrived furtively,' Maxwell recounts, 'for I expected to find Mij's pelt nailed out to dry somewhere ... for me it would have felt like finding the skin of a human friend, but I had to know ... I hope he was killed quickly, but I wish he had had one chance to use his teeth on his killer.'

When confronted, Big Angus took Maxwell to the spot where he had killed the otter, but was unwilling to admit it had been Mijbil, claiming it had been an old, mangy and emaciated otter. He covered his tracks, refusing to produce the body and stuck to his story, restating that the pelt had been too scabby to keep. Later a sympathetic neighbour who knew the truth took pity on Maxwell: 'I saw the body of the beast on the lorry when it stopped in the village, and there wasn't a hair out of place on the whole skin – except the head, which was all bashed in.'

It's hard to imagine how Maxwell overcame the feelings this revelation must have produced, but he did eventually replace Mijbil. His successor Edal survived almost ten years as a much-loved if trouble-some pet, but she too died tragically in the end, in January 1968 when fire destroyed Maxwell's croft. Maxwell escaped but Edal was trapped in her quarters and killed by the flames. Her charred body was lifted out and buried in a shallow grave by the rowan tree that grew nearby. Soon after, the house was razed to the ground, bringing the dream of Camusfeàrna to a close.

I take the ferry from Skye and cross back to the mainland to get to Camusfeàrna while the weather holds. I still haven't decided whether to go and see Jimmy Watt. It feels like an intrusion. But all of this is an intrusion. I know the place Maxwell describes is part fantasy; even the name is fake. He wanted to preserve the anonymity of the bay for as long as possible, but he knew that this would not be for ever. And now I, like many others before me, want to see it in its true dimensions and reassure myself that at least part of it *is* real. I am sure there will still be otters there, but this is not the only reason I am going. It's also to do with Maxwell's story. His friend Richard Frere sums up some of *Ring*'s attraction: 'there was a warmth in it that flowed in my veins like whisky. It did not seem to me like the work of the worldly, sophisticated man that I knew him to be, but like that of an incredibly literate child. It seemed that he had not lost sight of that "visionary gleam" which in most of us fades with the passing of the years.'

Curiosity about this childlike man gives way, as I near the real place, to a kind of internal map-reading; I'm travelling through the scenery I fell in love with in a book. Spending just one afternoon here does not feel like enough. When a place has so many layers of resonance it draws you in deep and it becomes hard to escape.

Sandaig is wild and remote. I must follow a steep, winding, uphill road from the village of Glenelg, find my way through an estate of Forestry Commission pinewoods, and then make a long walk downhill to the bay to find the exact spot of the lighthouse keeper's cottage where Maxwell lived. I worry about the weather turning; about how long it will take, and about being alone. The map makes it look like a long way on paper, and when I get to the start of the walk it is not well signposted. The green blocks on the map turn out to be plantations of massive pine trees which cover and hide any landmarks or contours. I am befuddled, and have forgotten the ferryman's instructions. Finally I admit defeat, find a house and ask. The house I choose, at Upper Sandaig, is the only one I can see, and turns out to be 'Tormor', where Maxwell's friends and nearest neighbours, the MacLeods, lived.

Armed with new confidence and proper directions, I move the van into a watery rut beside the track and set off into a wall of unhealthy-looking woods. Firs tower like scaly creatures on all sides, blocking out the light. The air is scented with Sitka spruce, resin and fallen needles. There is no wind, and hardly any sense of orientation. Eventually the planting opens out and I find a metal bridge over a gushing river. Silvery lichens droop from the trees like tattered clothing. Further on, light filters through, ferns appear and mosses green the forest floor. Although I am near the sea I still cannot hear it; there is only the faintest whirr of breeze combing through the pines. There is a strange, other-worldly creaking, as if a ship is floating somewhere on the other side of the trees. In a clearing, prehistoric ferns spurt like a green fountain through a patch of light. Outsized dragonflies flit about. In the grass I notice a strange grey-green fungus I have never seen before. Shaped like gigantic oak leaves, it sprouts and curls like mould over the path. The burn plunges its peat-brown rockiness into a gulley and I'm reminded of the instructions: 'Keep going down until you see the sea, then you're there.'

I remember a scene from the Disneyfied 1969 film of *Ring of Bright Water*, where the semi-comical figure of Maxwell, played by an endearing Bill Travers, arriving for the first time, is directed by a bus driver to the coast. 'You'll know you're there because, there's the cottage, and then there's the sea …' When finally he finds the cliffs and looks down upon the cottage, the musical soundtrack reaches a crescendo as the camera sweeps round in a panoramic shot of the bay in all its glory; we are instantly seduced by the view of mountains, white sand and glittering water.

There is no panorama here. The conifer woods hide the view and the midges are biting my eyelids. Patches of the forest have been shorn into rough grey stubble; the management of the woodland makes it feel like an entirely fabricated environment.

These forests were planted during the period after the Second World War when many demobilised servicemen needed a gentle activity such as planting trees. It was thought that this new woodland would provide timber and supplies of fuel for the ensuing Cold War.

They must have been dug in during the Fifties, just around the time of Maxwell's arrival at Sandaig. Now they have grown so high that they have not only altered the view but also the ecology and balance of the surrounding landscape.

The tapestry of peat bogs that were here has dried out. Previously, they had been growing for thousands of years. The sphagnum mosses created a surface supporting more than eight times their own weight in water. Tight clusters of moss formed a patchwork of hummocks blending greens, reds, pinks and oranges, like the formation of a reef or a rainforest. They held a vast range of microscopic plants and animals, provided food for insects, bog spiders and dragonflies. Under these trees, all that has gone now and the layers of peat and rotting moss cannot be replaced. The ground used to be covered in heather, bog asphodel, cotton grass, edible cranberry, bogbean and cloudberry. Over thousands of years layers of moss have rotted down to form peat, which held in and stored five thousand tons of carbon per hectare. When undisturbed it remained harmless, but the drainage of these carbon sponges has lowered the water table and led to vast amounts of carbon dioxide bubbling out of the bog.

When Maxwell first arrived in the spring of 1949 he might have felt he was walking into a pristine and intensely private landscape. The view of Sandaig bay with the backdrop of bright sea and mountains must have been overwhelming. Like the Romantic poets and many other writers before him, Maxwell was in pursuit of something untouched and untainted by modernity, and he felt he had found it here: 'I found myself on a bluff of heather and red bracken, looking down upon Camusfeàrna. The landscape and seascape that lay spread below me was of such beauty that I had no room for it all at once.'

Maxwell omitted to mention the people who had been here before, many of whom may have worked the land for generations, possibly since the time of hunter-gatherers. The area may have supported a population that, like the otters, found plenty of feeding on the bountiful shore. Nearby there are thousand-year-old remains, and ruins of ancient defensive brochs, tower-like castles built to protect families and their animals. Later, farming people settled in safety, or so they may have thought. A ruined barracks nearby housed Highlander-suppressing troops in the eighteenth century and, in the nineteenth century, many of the surviving locals would have been thrown out of their crofts during the Clearances, leaving little trace of their lives here.

When the hill begins to tip downward, I notice a whiteness creeping through the trees and with it a sound that must be water. Soon a green and welcoming stretch of grass appears, and a silver river that twists around – almost in a 'ring of bright water' – heading down into small waves that breathe at the shore. The grassy patch where the house once stood is eerily vacant. Not a single stone is visible, but the ground is bumpy as if there is something buried, and bracken has taken over a large part of the area around the site.

Maxwell had in fact moved into what had once been a well-established home; in the photographs in *Ring* it looks as if at one time the land around it had been cleared, managed and grazed. The lighthouse had long been automated, and the keeper's cottage and another croft nearby abandoned. An empty space remains, the edges overgrown with encroaching brambles, almost as if the place had never been inhabited. At the same time, having read about it all, I can feel the ghosts. In his book Maxwell described an old rowan tree close to the house which he superstitiously blamed for much of the tangle of darkness and bad luck he later encountered. There is no trace of either the tree or the darkness in this bright place; I can't help thinking that what he felt was darkness and bad luck might have been more to do with people than with anything supernatural.

Some of this 'bad luck' Maxwell attributed to his friend, the poet Kathleen Raine. She is barely mentioned in *Ring*, but therein lies a tale. Her literary output was prodigious, including fifteen volumes of poetry and an equal number in prose. In her autobiography she describes Maxwell as the love of her life, and reveals that much of her writing was inspired by her fixation on him. Crucially, she shows that she was pivotal in what happened to Mijbil at Camusfeàrna, for it was she who was looking after the pet otter when he was lost.

According to Raine, Maxwell was a deeply troubled man who sought solace and refuge in the wild beauty of the bay he chose as his home. In *Ring* Maxwell hardly speaks of Raine. He tells his version of the story through the filter of a charismatic persona, and we learn little about his true character and relationships. Douglas Botting, Maxwell's biographer, sheds some light on the bizarre and fateful relationship. When they met, he recounts, Raine was struck by Maxwell and felt she recognised a kindred spirit, and a vein of genius in him. When she found out that Maxwell came from the same area of the country, Northumberland, as her ancestors, she immediately formed an imaginary kinship with him and found herself falling deeply in love. In her writing she describes how they met again on many occasions. But he was homosexual, so her love was never reciprocated. For Raine, their shared feeling for the wild created a bond which was transcendental and pure, beyond any physical relationship. But it seems that Maxwell never acknowledged the depth of her love for him. Sometimes he kept her at a distance, at other times he would confide in her, providing a rocky basis for their relationship. In her autobiography Raine says she loved the otter Mijbil almost as much as Maxwell did, and describes being invited to stay while Maxwell was away so that she could look after him. Maxwell had left strict instructions about caring for the otter. He especially warned Raine not to let Mij roam too far, and not to allow him to swim out towards the local village. Anyone who has read the story or seen the film will be familiar with the tragic events that followed.

In the third volume of Kathleen Raine's autobiography she describes minutely the episode that led to Mij being lost. She poignantly tells how she made the fatal decision to allow Mij out of his usual harness. She didn't like the restriction of it, and was afraid it would get caught on something and that she would be unable to rescue him. What she didn't understand was that the harness would make Mij recognisable to others, and keep him from harm. As the otter disappeared off round the coast, directly towards the village of Glenelg, she feared she had made the most terrible mistake. Mij did not return. When it was clear he was never coming back, Raine described how it shattered her friendship with Maxwell. He did not return to the house for a year. Afterwards she said it was as if the two of them had been exiled from Eden.

Heartbroken at losing the otter, Raine describes her tears as 'streaming rivers of tears that would not stop'. Unable to console Maxwell, she held herself responsible for what had happened. Here their relationship becomes knotted in more complexity; Raine believed that some years earlier she had laid a curse on Maxwell. After an argument at Camusfeàrna, feeling rejected by Maxwell, she had called upon the mystical powers of the rowan tree near the house for justice: 'Let Gavin suffer, in this place, as I am suffering now,' she had cried. When he found out about this 'curse', Maxwell believed that it was the cause of what had happened to Mij. He made up his mind that all his misfortunes had been caused by her strange and dangerous occult powers.

Maxwell's friend John Lister-Kaye remembers Maxwell describing Kathleen Raine as 'that bloody witch'. She, on the other hand, wrote reams of startling love poetry inspired by their relationship. Her memories of her love for Maxwell had been 'like fixed stars', but on his final rejection of her, she wrote, 'the stars fell from heaven'. Even though the title of *Ring* is taken from a line in Raine's poem 'The Marriage of Psyche', and the poem is the frontispiece for the book, it appears without her name. 'He has married me with a ring, a ring of bright water', the poem goes,

Whose ripples travel from the heart of the sea,
He has married me with a ring of light, the glitter
Broadcast on the swift river.
He has married me with the sun's circle
Too dazzling to see, traced in summer sky.

The lines may be inspired by Maxwell, but as with many of Raine's poems, her love dissolves into an abiding universal spirit. Maxwell chose never to credit Raine for *Ring*'s title or for its frontispiece, and without their author, her words take on a transcendent quality. In her autobiography Raine reveals the extent of her love for Maxwell, but writes that he acknowledged the bond between them only once. It was written in Italian in a book given to her, in a dedication that suggested that they had met 'in the heart of an otter', as if this animal were the gatekeeper to their relationship. The final book in the *Ring* series, *Raven Seek Thy Brother*, contains more episodes from the rocky friendship between these two writers and shows that although Raine's love for Maxwell was destructive for both of them, she always remained devoted to him.

Their stories have faded, but the wildness of Sandaig remains overwhelmingly romantic. There are no people or roads here any more. The white sand is smooth and ribboned with water, a curlew's haunting call the only sound. Perhaps it was simply the soul that emanates from the ground and the spirit of isolation that brought Maxwell here.

The fictional name Camusfeàrna was chosen not only to disguise the location, but for its meaning in Gaelic: 'bay of alders'. These water-loving trees still grow thickly along the famous burn. A well-trodden path leads to where Maxwell's ashes have been buried, and a huge stone monument marks exactly where his writing desk once stood. Sculptural seashells, antlers, dried flowers and flotsam from the seashore have been laid as offerings. On the map it says 'Monument'; but this is more like a shrine. A massive boulder has also been laid as a monument to Edal, Maxwell's second otter, with the touching words 'whatever joy she gave to you, give back to nature' inscribed on it.

Visitors have placed an array of objects at the boulder – exactly the sort that would absorb an otter: a curiosity of seabird skulls, shells, pebbles and intricate bones. They seem like votive offerings, a collective apology, an outpouring of affection. It is evident from all the tokens that people come in pilgrimage in their hundreds each year. 'Animals,' Henry David Thoreau said, 'are all beasts of burden, in a sense, made to carry a portion of our thoughts.'

I leave Edal a smoothed pebble that I've carried from my home and make my way towards the burn that Maxwell felt was the soul of Camusfeàrna. This is the 'ring of bright water' that gave the book its name. I can hear it in its little gorge, as it tumbles and fans out into the bay, making an almost-circle of water around where the house would have been. The falls must be further upstream, hidden amongst what is now a thicket of alder, hazel and oak. Wild otters are surely attracted to this secret place, where, fifty years ago, the pet otters swam and played with their owner and his friends.

The river shoals like eels over pebbles. It is translucent as brown bottle-glass, making a crescent that curves elliptically with the sea-shore. Its trails have shifted over time, as if the water were constantly seeking new ways to join the sea. I follow its sinuous path toward the falls. Beside a tangle of roots in the bank is an otter slide, worn over time by wild otters as they enter the water. Otters follow the same ways, and perhaps this is one of the very places where Mij or Edal would have slipped in to play with eels and sea trout. Otters are creatures of habit and tradition and it's more than likely that the domestic otters sensed these ancestral ways. I find some tell-tale spraints, the otter's droppings, and some pad-prints on the curve of a silt-beach. Crossing the water I discover an old rope-bridge. Hanging hidden over the ravine of the stream, this bridge appears in Maxwell's book. I don't attempt to cross it, but scramble onwards through bracken, brambles, briars and low branches, until I am clinging to the edge of a crumbling precipice. I land feet first in shallow water, and at last I can see the falls.

The air is secluded. The water hefts into clouds of spume, and two rocks offer a platform on which to perch. The only thing to do seems to be to remove my clothes and give in to the frothing water below. I crouch down and put my fingers beneath the cold surface membrane of the water. I can see all the colours and shades of the stones that the falls have turned over and over, and with a quick intake of breath I immerse myself. The cold is sharp as needles. As the shock and pain give way to numbness, I notice the peaty water has painted my white skin weirdly bronze. I fight my natural buoyancy and dive down, running my hands over the migrating surfaces of pebbles.

There is nothing like a waterfall. For thousands of years this one has swelled with cloudbursts and dwindled with the sun; the extrusion of granite that forms its towering cliff pours hundreds of tons of water in a ceaseless churn. Inside its green magic, my body is held in strange weightlessness, transformed into part of the liquid suspension of silt and alluvium. Beneath me is the pool's coveted floor – a mineral hoard of pebble-survivors, each one a traveller through unfathomable strata and millennia. Other layers of resonance imbue this wild swim; I know that Maxwell and his friends, and perhaps even more importantly his otters, all swam here before me. The memories imprinted here in the rock and the trees, and all the stories I have heard and read, combine to charge this place with a powerful spirit.

I resurface with a trophy, inhaling gulps of warm air. This flat piece of shale will travel all the way home with me. I know its shiny surface will fade, and it will look ordinary on my shelf; later I might release it into new waters where it can continue its journey.

The air is charged with something else when I get out; the midges are massing in an ominous afternoon ritual. I do believe they have been waiting for their moment. Before I can get dry they make a start. How can something so tiny cause such damage? At the first few bites I feel a moment of panic, and try to squeeze into my clothes while I'm still sticky with water. Trapped inside a trouser leg that seems never to have fitted me, my foot jams fast. I slip, bang my elbow, lose a boot into the water. There follows a tangled pantomime of wriggling and cursing. Even when I am dressed, the midges continue their relentless feast, and make a meal of my face and neck. I have forgotten to pack both repellent and midge hood. Smarting and dishevelled, I limp out into the midge-free breeze of the rocks and beach.

It is low tide. I lay out my sodden boot to dry and lie down on the sandy incline, hands behind my head, to gaze at the view. All the islands can be seen from here. The tall peaks of Skye, and in the distance the uneven backs of Rhum and Eigg. Further south are the magnificent roadless contours of Knoydart.

All around, close-to, are swathes and scatterings of cockle, cowrie, winkle and limpet shells. I pick up an interesting quartz-veined pebble and the broken-open whorl of whelk, its intricate listening ear exposed to the air. The mussel shells are worn to a thinness of lilac and pearl. I get up and quarter the beach, walking to and fro foraging for treasure. From a swirl of sand and weed comes a perfect razorbill skull, its wedge of beak like black plastic. Further on, the white bone of a gannet's head, complete with beak, neck vertebrae and mohican of cream feathers. Then, my best find: peering out of a clump of dessicating bladderwrack, what looks like the eye of a sea-sculpted otter. With one pull, out comes a piece of driftwood eroded into the luxuriant shape of a dive; water and salt have worked fur into its sides and left worn nodes where the front feet would fold as it swims.

My knapsack, bag and pockets bulging, I make a last search for the foundations of the house. It has not even left a shadow. I stand for some moments listening, imagining the walls, how permanent they must have seemed. At the memorial stone, the exact spot of the desk where *Ring* was written, I imagine I see Maxwell going about his business in the evening light, closing a door, smoking, calling his otters or their keepers. The chaos of their life together comes to mind. There were huge cooked breakfasts of bacon and eggs, chunks of bread, thick porridge and steaming coffee, Terry Nutkins remembers, but little structure to their days outside this, with few formal meal times. He remembers Maxwell disappearing off into his study for hours, forgetting that he had promised to provide an education for his wards. It was often lonely and as the ferries passed in the Sound, full of people, the boys watched from the beach and wished for the comfort of more company.

On my way back to the van, the sky darkens, the wind gets up and a squall of rain drenches me right through. Real Highland weather. I hurry into the sheltering woods. The pines shiver and creak, their voices punctuated by the irregular patter of loosened cones. In a glade of giant ferns, the carcass of a fallen tree protrudes, its branches like ribs dripping with lichen.

As the trees thin out, I spy the van and run to it with my bag of finds. Climbing gratefully inside, I set off for a new camping spot. More winding roads and driving, this time in very clammy clothes. When I reach the village my courage wavers. There is an inn at Glenelg. I stop, reverse back to it and pull in. Perhaps I could allow myself just one night of luxury. Hesitating in the car park before going in to ask about a room, I tell myself that were it not for the midges, I would sleep out. It's that uncanny knack they have of insinuating themselves in their thousands into every available gap in tent cloth or vehicle. My midriff and face are still swelling as a result of today's swim and the previous night's unforgettable wild-camp where I discovered the full horror of midges in still air. I know now that on August evenings every millimetre of uncovered human flesh is at risk: ankles, calves, neck, face, lips, nostrils, ears and scalp … resistance is useless, and no bare part is spared.

'How much for a room?' I ask the Irish gentleman behind the bar. He looks me up and down, observing with admiration my recent battle scars. After a pause and some mystifying sniggering from the row of punters at the bar, he says, 'Just give me a minute,' and disappears. I am left at the mercy of the locals. 'Don't worry,' confides one, swaying sympathetically closer, 'you're in good hands with Kevin.' Before I have time to think of an adequate response Kevin is back with a key, which he slaps on the bar like a bet. '25 quid?' There is more amusement from the row of onlookers, and we set off to the sleeping quarters. I smuggle my bagful of sandy trophies up the stairs. 'Will you be wanting supper?' Kevin enquires. I don't fancy an evening of more sniggering. 'I bought some food already,' I answer, rustling my bag. Kevin lingers at the door; his eyes are moving over the contours of my gannet bag. A sharp beak is protruding, along with outlines suggesting other body parts. He hands me the key and moves back out into the corridor.

Safe in my room, socks hung up to dry, I unpack my treasures. First I soak the gannet head in the sink. It will need a few hours to dissolve the smell. Then I put the bath on, stretch out on the bed, and over a hearty chunk of cheese and an apple, set about admiring the objects I

58

have picked up. Arranged on the smooth surface of the counterpane, a little space between each, they seem like a set of haiku poems. A worn whelk shell, a driftwood otter, shale from the waterfall, a veined pebble and a smooth grey goose feather, which I found on the space left by Maxwell's croft house. It is just the size and shape of a writing quill.

In the morning, washed and brushed, I decide that I will go and see Jimmy. He lives in a white house perched on the rocks above the sea. The house has been lovingly restored. It's a poignant domestic scene amidst the remoteness of the setting. There is a gravel drive, a well-tended garden, and parked alongside it a neat little wooden yacht.

I take a deep breath and knock. A friendly-faced man opens the door. His hair is white, but he is tall, sturdy and ruggedly handsome. It is unmistakably Jimmy. I hold out the book with his picture on the front and explain myself. After a moment's thought he lets me in. Once inside, seated at the large kitchen table, Jimmy is extremely hospitable, and does not seem to mind talking to me. There is a deeply rooted homeliness here. An Aga warms the room, and a black and white collie sleeps at the hearth. We drink coffee, and while we talk, homemade brownies are produced. I ask Jimmy to tell me about the otters. 'Edal was a big otter, with this amazing skin,' he remembers enthusiastically, 'and too much fur, almost as if it didn't fit. You could stretch it right over her head like a hood,' he indicates, with a two-handed motion that implies a big stretchy jumper.

59

Jimmy describes being trusted to look after everything, not just the otters, while Maxwell was busy with writing or travelling. He was at Sandaig for eight years and so closely involved in all Maxwell's doings that he became his heir. During that time Maxwell had depended on him to oversee the running of the house, in good times and bad. In *Ring* it is the young Jimmy who often saves the day whether there's an otter crisis, or indeed any other drama. In *The Rocks Remain*, it is Jimmy who one stormy night leads Maxwell to shore from their wrecked boat. He teaches a gaggle of unfledged geese to fly, repairs holes in keels, plucks a spitting wildcat from where it has been lodged for several days up the chimney, and cares for the temperamental otter Edal. Once, when an impetuous tantrum found her bolting far from the croft, it was Jimmy who carried her back home, draped confidently around his neck like a fur collar.

It was clearly an extraordinary life. 'At first there was no electricity, nor any telephone, and at times life was cold, and harsh. But there were many happy moments,' Jimmy tells me. They used to fish in the burn and play on the string of islands adjoining the bay, and it was from here that they watched their adopted family of wild goslings fly away. He writes about the return of these geese in his foreword to the new edition of *Ring*, and when I get the book out for him to sign, he opens it up and looks at it in surprise and wonder, rereading the passage as if he had forgotten it: 'We heard the call of the geese high in the clear autumn air and called up "chck chck chck" at the tops of our voices; and the geese we had reared, and who had flown away in the spring, circled and lost height by dipping a wing as geese do, and with

60

a huge clamour of greeting landed in the sand at our feet. It was a moment of overflowing joy that these wild creatures had chosen to return.'

The story Maxwell made of his life – the adventures, shipwrecks, car accidents, financial disasters, fires, catastrophic love-affairs, and finally the tragic illness and demise – couldn't have been more dramatic if it had been invented. But nevertheless, he left behind a story of a human search for connectedness to a place and to the wild. Jimmy Watt's relationship with Maxwell seems like that of father and son, and they obviously had a strong mutual understanding. Jimmy remembers Maxwell as a solid friend who adopted him as part of the family. Although he does not talk about Maxwell's death, something about the loss is visible behind his eyes.

Jimmy is much happier talking about the wild otters that he sees regularly from the rocks around his house, and wants to show me a film taken on his camera that morning. We watch a family of three, mother and cubs, frolicking on the rocks, fishing and squabbling over prey. They dive, resurface, play and bicker, ignoring the observer. Jimmy is a pro; he knows instinctively how to stalk and be still around otters. It seems that his time with Maxwell has instilled in him some indestructible connection with and love of wild nature. How could he not settle here? The house he lives in today is not far from Sandaig, and looks down the Sound towards it: 'On stormy days when the ebb tide meets the south-west wind,' he wrote in the foreword to *Ring*, 'I can look through the spray to the bright white sand where the geese landed that memorable day.'

Does he often go back? I ask. 'I was there yesterday. We sailed down, and it was lovely – completely empty.' I nod sagely, feeling a little relieved, no, very relieved, that we didn't bump into one another during my clumsy foray in and out of the burn. 'It's so overgrown now, as if nobody goes any more,' he tells me. I nod into my note-book, and ask about the trees of the conifer plantation: was any of it like this before? 'No, the conifer trees have been devastating for the environment. The water in the burn used to teem with life. There were not as many trees before. It used to be more open – just moor-land and heather. It was beautiful, you could see all around.'

Maxwell died of cancer in September 1969 and was not to know the continuing story of the wild otter and its decline. But ever since the first book, thousands of people have come in a steady trickle to Sandaig. Awareness spread, otter hunting was banned in 1978 and, where otters had disappeared, wetland habitats were painstakingly studied and restored, farming methods were altered and the use of chemicals toxic to the ecosystem reduced. Although it is not measur-able, writers like Gavin Maxwell, working quietly before and alongside the scientific research, must have played their part in this awakening.

Driving homeward through the wilds of Glen Shiel, beneath the ice-sharpened peaks of the Five Sisters of Kintail, I stop to let the engine cool, and think about other forces that shaped this landscape. The fire and ice seem tangible, the millennia of meteorological and geological forces manifest. There are more ancient, elemental battlegrounds than the human ones here. The river valley gouges its way down through smoothed rock, and the stream is stained peat-dark. I peer into it, examining the muscular sides of the gulley, and wonder about otters up here. How many millions of generations have passed through this spate water on their journey from the river to the sea? The water riffles with a dark intensity, connecting the source high in the rocks to the tumbling ocean. I walk over to where I camped out two nights previously and find the depression that my small tent had made beside the thick woods. I muse that apart from the voracious insect life, any wild animal could have come nosing round my tent in the small hours while I quivered inside. With only the nylon micro-fabric of the fly sheet for protection, some ancestral fear always comes back, especially in the dark. There are marks where an unidentifiable creature had meandered over the grass; there are pauses and scuffles, and places where earth has been turned over in a small frenzy of feeding. There may be no real danger, but being alone like this is all about learning to be comfortable in the wild.

Just as this thought crosses my mind, a feeling of not being entirely alone creeps up on me. I detect a shadow, and turn my head. As with my first otter sighting, a voice inside me says *it can't be*. With a cold trickle of recognition, I see something huge moving out of the trees towards me. It materialises into a bristling brown body rippling with muscle and a tremendous head weighted with hefty chins. *Wild boar*.

Its prehistoric-sized ears point at me like radars, its eyes are black with menace. My heart flips and tears push out of my eyes. I have interrupted its foraging, and with four trotters planted in the peat, it is making one thing clear. It is not impressed at being disturbed. I have just enough time to take in the hairy snout and the cutlass tusks, when the air is shaken by its baritone grunt. My legs are reduced to plasticine. What do I do? Run? Stare it out?

Nobody had said anything about boar in this area. I had only heard about them spreading in other places, much further south, where having burst out of captivity, they quickly revert to a feral state. When this happens, a boar can be impossible to recapture. This brute may have been wild for years, growing freely to its current dimensions.

Another threatening grunt, as if I didn't hear it the first time. And now, remembering how bellicose they can be when cornered, how, if surprised, they might charge, my feet act before I can think. They fly over slippery leaf-mould and fallen logs, over the cold rush of the river, through brambles, over rocks and branches, and finally up to the familiar comfort of the tarmac.

I don't look back until slammed inside the safety and protection of my van. Chest bursting, I try to regain control of my breath, then scan my trail. I see only stream, grass and woods. In the dark between the trees there is an almost imperceptible twitch of leaves, and nothing more.

Watershed

A ripple firms
into a jink of fur,
undulates into a clay-coloured pelt
as he rides the undertow, steers through
the backwash, burrows into the depths.
If you wait and watch for his return
you'll hear the river's many voices utter his name.

Rebecca Gethin, 'Fluent'

Bedridden with autumn flu, my eyes dawdle over nearby shelves and stop at the old copy of *Tarka the Otter*. Inside the front cover, my childish handwriting claims ownership of the book in blue felt-tip pen, and a fading Puffin Club sticker reiterates the fact. My legs are too wobbly to go out ottering at the river near my home, and this book is the next best thing. I blow the dust from the book's edges and sniff the aroma of the paper. It hasn't been reread for years.

The opening passage of *Tarka* describes a nook of England that is not far from my home. Henry Williamson saturates his descriptions of the North Devon landscape with reverence: 'Twilight upon meadow and water, the eve-star shining above the hill, and Old Nog the heron crying kra-a-ark! as his slow wings carried him down to the estuary.'

Straight away I'm back with an old love, drawn inside a rhythm written directly out of the seasons: 'Yellow from ash and elm and willow, buff from oak, rusty brown from the chestnut, scarlet from bramble – the water bore away the first coloured leaves of the year.'

Williamson's eye is at otter-level and every contour is under the magnifying lens. He notes everything, from the weather and the season to the insect and bird life. As I reread *Tarka*, I realise that it is a book that is hard to categorise. It's not fact and it's not fiction; it wanders through details of the life of an otter; it maps the memories of the landscape, it daydreams: 'Over the old year's leaves the vapour moved, silent and wan, the wraith of waters once filling the ancient wide river bed – men say that the sea's tides covered all this land, when the Roman galleys drifted under the hills.' The story meanders through time, invoking all the layers of hidden and familiar places around the water.

Above all, the description when we meet the otters is sibilant with sounds of the river: 'The rising sun silvered the mist … she was young, calling to the dog with a soft flute-like whistle … the song of the river hastening around Willow Island stole into the holt and soothed her.' The narrative observes every detail of the water-scape, and through its creatures and cycles Williamson tells a tale of a whole web of life. A self-taught naturalist, as well as being an expert tracker of otters, he writes with such precision and tenderness that I can't stop reading.

Williamson was born in 1895 in London and began writing nature diaries as a young man, but it wasn't until after the First World War when he moved to the Devon countryside that his writing career began to flourish. He wrote articles and several novels, and the book he is most famous for, *Tarka*, was published in 1927. *Tarka* was a striking and realistic story of the life of an otter in North Devon, and it was an instantaneous hit. It had long been a best-selling classic by the time he died fifty years later, in 1977. What I know now, but didn't realise the first time round, was that simply by calling us to share in the vulnerable, intricate beauty of the otter's world, Williamson was also asking for its preservation. I had no idea either, as a child reader, that while Williamson was writing this delightful story, he was about to formulate some unpopular political ideals. His friend Ted Hughes later suggested that perhaps these were born out of a longing for the pure, vital energy that Williamson felt certain places had. In *Tarka* we only see his love of the English landscape and its history, and nothing of the pronouncements that would later bring him conflict. He seems to long for something that has been lost or forgotten in the land, but there is nothing to discredit the writing. It was only later that political ideas seeped into his prose and some of his books fell from favour.

In *Tarka* the innocence of the description of the landscape's delicate system has a seductive quality. I had so loved his story as a child that in adulthood I moved to live in the South West, and made my home in Devon, hoping that in this greener, wilder, wetter place I would be closer to otters. But after all these years, even though I know they are here, I still haven't seen an otter outside Scotland.

I console myself with the book as if it's a talisman. Rereading *Tarka* is hypnotic. Several pages are devoted to the life cycle of the eels in the river Tarka inhabits, and I feel my own gills growing as the words transmit all the river's slimy, watery detail. A series of seven silver bubbles breaks the surface of the water and materialises into an otter, and our journey continues at exactly otter height – drawing us just above, and sometimes below, water level. A damselfly alights on a leaf; an owl floats like mist over the river; reeds rustle. Everything is magnified by the viewfinder of the otter's senses. Williamson obviously spent a large part of his life listening to and watching his patch. It makes me wonder: how much do I really look at mine?

The book's fusion of sense and nature seems to slow down the crazy merry-go-round of the Earth. All that coming and going; I need to stop and be in one place for a while. I look out at the birds on the birch tree closest to the window with a new eye: one female wagtail came to eat from my balcony all summer. I watch the starlings and, with a pair of binoculars, peer at the constellations on their plumage.

The flu drags on into some strange post-viral illness. I can't watch the News; it's too upsetting. Any amount of noise makes me jump. There is no longer a protective membrane between me and the outside world. 'You're depleted,' diagnoses one friend wisely. 'No more trips to Scotland. Just stay put and do nothing.'

Every so often, I hear a commotion in the branches of the tree outside my window. Sometimes it's the sparrowhawk coming through. A stealth hunter, she sweeps in and knocks her unsuspecting victim from a branch. Each time I have to heave myself up and look.

My cat is the most active predator. She enters through the cat flap, loudly announcing live trophies through mouthfuls of feathers. Sometimes, when I can, I corner her and pick her up by the scruff of the neck till she drops her prey. If it is still alive, I place it somewhere quiet where it can recover. Once I watch a tiny goldcrest revive in the palm of my hand. A series of goldcrests follow, and one by one I try to nurse them back to life. Not all of them make it. These tiny birds may have come from somewhere hundreds of miles away; they could have survived vast stretches of North Sea, battled with high winds and storms, or come down from the moors to pass the winter in a gentler climate. And when they get here they face bared teeth and cutting claws.

Another day it's a song thrush. The cat is removing breast feathers by the time I get there. As I prise the bird away it glares at me and fights back. I experience the needle-sharp beak, curved like a sabre for butchering snails. Something about this wild songbird using its equipment as weapons against me wakes me up. Where has it come from? Where will it go?

We don't pay enough attention to this tribe that lives alongside us. They yield to us because they have to, and endure the relentless fall-out from our lives. The chemicals that magnified through the food chain and wiped out apex predators like otters and peregrine falcons are no longer used, but I suddenly think of how many times I have eliminated my slug and snail population with pellets that are probably toxic to the whole ecosystem. What horrors are in my concentrated washing-up liquid and my cleaning products, and where are they ending up?

This moment is what the American writer Henry David Thoreau called a 'return to the senses'; a common, unifying epiphany among lovers of nature where time can seem to stand still. Annie Dillard, in *Pilgrim at Tinker Creek*, records when it happened to her. She passes a tree near her house; one minute it is a normal tree, then suddenly it is aglow with colour and light. Later she realises: 'the most I can do is try to put myself in the path of its beam'. She describes the experience of this clarity as 'less like seeing, than being knocked breathless by a powerful glance'. As the colours on her tree die down, she is left ringing with energy: 'I had been my whole life a bell, and never knew it until at that moment, I was lifted and struck.'

By the end of September I'm feeling stronger and, still ringing with Annie Dillard's words, I venture out to look at the 'free fringed tangle' that Dillard says begins just outside our gardens. Beyond this, anything could happen. I head for my own Tinker Creek, a muddy tidal arm of the river that runs down the valley not far from my house. Walking along the rugged green lane, I begin to feel better. The same creatures are living in these hedges as in my garden at home, only here they are more sensibly organised. Water erosion has created a sheltered microclimate for the elaborate strata of life that envelop the sides of the lane. Around me is a system as rich, complex and intelligent as a rainforest. There is a dark under-storey, and above it layers that confuse in a maze of prickles, stings, fronds, leaves, twigs and buds. It buzzes and dazzles with life; I cannot make sense of it, but I don't need to. It has a sense all its own.

The lane winds through a mire of Devon clay all the way down to the river. My boots are engaged in a ritual bonding of feet and earth, and soon cake with a substance so sticky that I feel centimetres taller. Hazel, ash and sweet chestnut form a fidgeting roof above my head. I look through the hedges at the fields beyond. They are bland as Astro-

Turf. Birds should thrive in this mixed patchwork of fields and woods, but these hedges are stranded in a monoculture organised for the production of cattle.

At the tidal flats where the creek meets the river proper, the path disappears and my nostrils fill with the tang of brackish water and seaweed. The mud is doodled all over with aimless wanderings; tiny, scribbled bird prints; erratic circlings of some small mammal. Further on it's pimpled with a million worm casts, and I can see where a fox has been trotting on sticky paws to the water. Many of the tracks seem not to have a clear will to get anywhere in particular. My eyes wander, distracted by fallen hazelnuts, bright red splashes of hawthorn and fronds of rosehips. The fruits of *Rosa canina*. English dog rose. I pick off a ripe hip, bite into it and suddenly an aroma of spicy syrup floods my memory. We used to have it in winter to fend off sore throats: Delrosa Rosehip Syrup. I can remember the glow of amber in the bottle and the sweetness of each spoonful.

I climb the bank and scale a small tree to prise the dangling rosehips away from their thorns. Some are overripe and burst as I gather handfuls to bring home. From my vantage point I pause. I have a bird's-eye view of mist creeping in; shadows are starting to lengthen and everything is wilting into a palette of deeper colours. My gaze is drawn over the slippery flanks of the creek, and in the dimming light my senses reconfigure. Moths are coming out, attracted by vaporous messages that my own nose can hardly detect. In a last sliver of reflected light, something on the water distracts me. It's moving like an animal but made out of liquid. It ripples for a moment and leaves the

73

hint of a wake. A long mud-brown slither slowly becomes more creature than branch. I see a smooth head; the contours of a brown face
with ears, whiskers and the dark holes of two nostrils flowing pur-
posefully downstream. The barely perceptible bump of its dive and a
lingering tail-tip convince me.

An otter. So strange and subtle that I could almost have imagined it.

Autumn has taken hold of the trees. For several weeks I have been
coming back to the water's edge. On some parts of the river I'm tres-
passing, drawn along its length by the sight of that otter. It's my
home patch, but because this stretch is private land, nobody else ever
comes here. No dog walkers, no children and apparently no fishermen.

At this time of year, the river Dart in Devon is at its most beautiful.
October birch leaves flit gently down and are taken on the membrane
of the river surface. Banded demoiselles flicker their electric bodies,
fire-blue tubes on glassy wings. The water moves over roots, curls in a
bend into the weir pool. I stand amongst water mint and tall grass,
my nostrils filling with the tang of mushrooms and mulch. On the
opposite bank, pale tree trunks reflect in the water surface. A raft of
interlocking oak and beech leaves drifts by, bending the water with its
weight. An otter could live its whole life in this patch and never encounter anyone.

I feel rooted here, but at the same time I am an outsider. My scent
must appear complicated, or even hostile, to the creatures whose

home this is. I do not expect to see otter; it will sense me a long time before I am aware of it.

I have found the otter's spraint – its small, oily droppings – and some irregular five-toed prints in the silt. The spraint are fresh, tubular and greenish-black. When I look more closely I can see that they're alongside others that are greying and desiccating in the sun. Otters learn from one another where to leave their signs, and it's possible that they could have been visiting this spot for generations.

The otter-hunters used to call these signs 'coke' because of their black, ashy appearance. They contain all the ecology of the animal; otter spraint is a language all its own, scented with ancient otter vocabulary. This dropping points downstream, placed on flat rock by the water, above where the flow folds into the gushing weir. The otter must have exited to avoid the fast-flowing water and deposited her signal here. I lean over it as if it's a manuscript which is only partially decipherable. It is inky, and still sticky, which means it is recent. In it I can see bits of fine fish bone, scales, vertebrae and what look like fragments of a shiny beetle case. Look at any of these under a microscope and you might identify roach scales, pike vertebrae, bullhead bones, but what it says to another otter I can't ever be sure.

Most noticeable to me is the spraint's aromatic smell, like lavender or jasmine tea. Otters are the only animal I have come across whose excretions smell like perfume. I now know that they leave it in a trail for other otters to find, and it can be followed by humans too if you know what to look for.

Close-by is a tangled knot of dried grass. I imagine the otter crafting it. Some call them 'castlings', and it is thought otters leave these woven sculptures to mark their territory or make their spraint stand out. A spraint might be stuck to the very tip of the grass to maximise the chances of another otter finding it. They do the same thing on river beaches, making small scraped-up sandcastles on the exposed silt or gravel banks. Other otters will come across them, sensing layers of meaning in the scent on the grass or sand. When the otter was hunted regularly, hounds were trained to recognise and follow these trails, and they could even find the otter's fresh scent as it lay on the water; this was called the 'drag'. My own dog never seems to notice otter scent. He is not a hunter, and when out walking with me is thoroughly absorbed in reading the fascinating layers of other smells along the bank.

We humans pride ourselves on being able to see in full technicolour, while many animals look out at a monochrome world. What we forget is the extent of their other senses. Most other mammals smell the world, and hear it, in 3D. Through his nose, the domestic dog experiences dimensions unimaginable to us; he has more than 220 million olfactory receptors in his nose, while we have a mere five million. Wild animals have senses that are even more powerful. Their hearing is so acute that they can detect sounds and frequencies of which we are totally unaware. Well before I have reached the water's edge, the otter will have made itself invisible.

Leading up the bank from some tall reeds is an otter path. It spans one otter width and travels from the water, over the sand, across grass, through a bramble patch to a holt or den hidden in the dark tangle of an ash tree's roots. There is a great deal of spraint around; I can smell the holt, which might be marked each time the otter enters or leaves. Sometimes the odour from these can be detected from quite a way off, but as it's against the law to approach an otter holt I must avoid going too close.

Ash trees are most popular with otters because their roots form a complicated system of shelter below ground, and are often right by or even overhanging the water, so that the otter can slip subtly in and out. An otter may also sleep on a rocky ledge or tucked away in the reeds. To enter the water they prefer to use points where there is cover – branches or undergrowth – to increase stealth and invisibility. They are good climbers and will sometimes climb up into the hidden shelter of a pollarded willow to sleep. During the day, people, dogs and cattle may walk past unaware that an otter is curled up right above their heads, fast asleep, protected from any disturbance.

In the shallows, the river is inscrutable. The otter's map of this area must look very different from how it does to me. The otter can see a short distance underwater, and the curvature of the lens of its eye can be adjusted to improve vision. From below, the water riffles with potential. Under the surface is everything that is important. The upside-down contours that an otter sees must seem vivid and alive. Everything is in a constant state of transformation: algae and ribbons of weed form shifting curtains; invisible currents bring hints of eel and stickleback; pebbles roll away on their journey downstream, forming piles and eddies that create endlessly changing feeding grounds. Here nothing is ever the same. Occasionally, dark shapes course upstream like torpedoes; an otter's senses can detect these instantly. A larger fish is a lucky find; generally otters will predate upon anything from tiny crustaceans beneath rocks to larger crayfish, mussels, snails and amphibians. The muffled chorus of sound created by all this movement, the rustling, ticking, shifting resonances that an otter experiences under the surface of a river, must sound like a chaotic and discordant orchestra. The flicker of colour, light and shadow must present an additional challenge. When otters' eyesight varies according to the amount of light or murk in the water, they can rely more on their super-sensitive whiskers to pick up minute movements and vibrations from their prey.

On the banks there is more spraint scraped into heaps. If it is a male, it could be transient; males have to disperse to establish a range that is safe from other males, and are more nomadic than females, who tend to stay on one patch of river. If a dog otter locates a bitch in season they will mate, and then he might move on to find other females. Couples do not stay together for long after mating, although the dog otter may occasionally pass through and visit. The female may see this as an encroachment, and he may not be welcome; mates are often quickly chased away by protective or territorial mothers. The female may choose to give birth hidden away from the male's usual patrol. A dog otter can be aggressive, and if he is an outsider, may try to kill cubs. This would mean that the female would soon be ready to mate again, and the new male could return to ensure the continuation of his own DNA.

Otters are sensitive to environmental change, and may give birth to fewer cubs when food is in short supply. In much of mainland Britain two or three cubs is the usual number, but occasionally four can be born. With a survival rate of about 50 per cent, maybe only one cub will reach adulthood. Even with this low number of survivors, young cubs are sometimes deliberately abandoned, perhaps because of ill-health, a birth defect in the cub or a variable food supply.

Some species of otter, for example the North American river otter, *Lutra canadensis*, have a mechanism where, after mating, the fertilised embryos can be held in the uterus but not implanted for up to eight months. The delayed implantation could be a survival device because of the harsher climate in North America or some other difference in the ecosystem; whatever it is, our British otter does not appear to possess the same ability. Gestation in our otter is 60–63 days, and it is thought that the dog otter leaves cub-rearing entirely to the female. He may guard up to three females in his range, so he is constantly busy and on the move, shifting his position to vary foraging and op-timise feeding, and to chase away other males.

It is thought that courtship usually takes place underwater. Scien-tists have observed this pattern of behaviour, but with such an elusive subject it is easy to make assumptions. We may have only ever seen the otter mating underwater, but why shouldn't it happen elsewhere? Henry Williamson, who was an amateur naturalist rather than a sci-entist, was also an acute observer of the otter's behaviour. He corrobo-rates the underwater-mating testimony, but we don't know whether he was fictionalising, or basing incidents in his story on what he might have read, or on what he actually saw. In *Tarka*, his guesses might have been as close to the truth as those of a scientist, as when Tarka's mother meets and plays with her mate: 'Tarka dived to see why this strange thing had happened, but the dog turned in a swirl of water to snap at his head ... He saw her swimming with her head out of the water, with the strange dog behind pretending to bite her. She was heedless of her cub's cries and dived with the dog in play.'

In the documentary *My Halcyon River*, the naturalist and film-maker Charlie Hamilton James captured exactly this type of behaviour on camera. Charlie followed a mother otter and her two cubs living near his home. He became so familiar with them that they lost some of their natural suspicion and he was able to get quite close. He filmed the moment the family were visited by the cubs' father. In one film clip, shot at night-time, the mother greets the dog otter, while the confused cubs call in distress. Sensing her cubs are big enough to be left, she ignores their cries and disappears for a night with her mate. The following year, Charlie notices that the grown cubs have dispersed and the mother reappears with a new litter.

In my search for the otter I have begun walking further along the river that flows through my home town. The river Dart has its source high on Dartmoor, in a boggy marsh near Cranmere Pool, eight miles from any road. Dartmoor is southern England's wildest place, and it covers over 370 square miles. There must be huge amounts of ground-water hidden in the bedrock of this rugged granite outcrop, as twenty rivers spring from here. Many of them, including the Dart, are torrent

rivers that rise quickly after rain. Over time these have carved themselves steep gorges as they rush off the sides of the moor. Dartmoor is
not true wilderness, but there are very few roads and the landscape
hints at the wildness that was once here.

Dartmoor's unusual geology is visible in the spectacular formations
called tors which protrude and sprout like toppling art installations
from almost every high hilltop. The formation of these has always
fascinated me. The moor is a granite batholith which dates from the
Carboniferous period, and the tors were begun when hot magma was
injected up through the earth's crust. The molten magma cooled and
as it contracted, lines of weakness formed. When the sandstone and
slate that lay on top of the granite were eroded, pressure was released
and the granite expanded. The horizontal weaknesses were eroded
over time by the weather, producing smooth folds and creases which
create the impression that the tors are sculptures made from tremendous boulders, piled teetering on top of one another.

Surrounding the tors, Dartmoor is covered with layers of peat bog,
with patches of oak and beech woodland in the valleys and around
the steep edges of the moor. The highest point on the moor, High
Willhays, is a tor over 2,000 feet above sea level. It is often snow-
capped in winter and at this time of year it is starkly beautiful: the
greys of granite tors and sparse dry stone walls contrast with russet
bracken-covered moorland punctuated by grazing herds of half-wild
ponies. The slopes are washed with metal hues of gushing rivers and
empty, wind-blown woodlands.

Around Cranmere Pool, the rivers Dart, Taw, Teign and Okement are born. It's a confusion of sources; pools are hidden in a wide expanse of soaking moss and sinkholes. After rain, water tips over the tops of your wellingtons. This high moor is nothing but blanket bog, granite outcrops, peat shelves, marshland and fathomless springs. Here and there the oily sump of peat stains the water bronze. There is no visible sign of otter up here, but Henry Williamson thought they were present. He walked all the way up the river Taw – the river where Tarka lived in North Devon – looking for otters. Crossing the watershed, he went to Cranmere Pool and claimed he found otter signs nearby. A low ridge separates the source of the Taw from the source of the Dart, and an otter might easily cross the short distance from one river to another.

This desolate area takes half a day to get to. It is one of the most quiet and hidden parts of the moor, and is only for the hardiest and most fearless of walkers. The watershed goes in every direction. The Teign and the Dart form wooded slopes to the east and south, with the Taw and Okement's rocky gorges to the north and west. My navigation skills are not excellent and being here always makes me jittery. Mist can fall without warning and the marsh will pull off rubber boots at will. Deep peat trenches loom up unexpectedly, and it's possible to fall into one up to your chest. I move carefully, looking for the young rivers that spring up and make their way through the peat. This is blanket bog: a mantle of peat one metre or more in depth. It is the most southerly bog in Britain, and there is very little further south in Europe. Fragile sphagnum mosses, cotton grass and rare carnivorous sundews grow here, and the matted vegetation forms a living protective skin over the spongy peat.

Williamson came to the moor not only to track otters. He came to reconnect with the wild. As a boy he had been an avid nature-watcher and diarist, but in 1914, when he was just nineteen, he enlisted in the Army and went to fight in the trenches of the First World War. 'For two days and nights we have been in nearly 36 inches of mud and water,' he wrote in a letter home to his mother in December 1914: 'Can you picture us sleeping standing up, cold and wet half way up our thighs, and covered with mud. As we crept into the trenches at dead of night, the Germans sent up magnesium flares, and we had to crouch flat while scores of bullets spattered among us.'

He was 'invalided out' several times, and in his letters home describes his nerves as being 'a bit joggy'. When he returned after the war, the wilds of rural North Devon must have seemed like heaven. He described his memorable journey there. It was March, and he had set out by motorbike on a fourteen-hour trek from London, without an overcoat. 'How it rained,' Williamson remembered. 'But I hadn't cared for the wet or the mud, I was free, this wasn't the rain and the mud of the Somme or Passchendaele. Grand feeling, causing one to shout and sing as the wet trickled down the small of the back and around my stomach.'

Drawn to the fresh air, space and the solace of the open sky, Williamson saw Devon as a place to recover, and as somewhere to recharge his writing career. He rented a tiny thatched cottage next to the church in the village of Georgeham. Barn owls nested in the roof of his new home, and a stream ran behind and around it.

Once settled, he started to chronicle rural village life – all that he saw, from plant life and birdsong to the local inhabitants and their soft North Devon burr. A quirky dialect comes through in some of his writing, which outsiders found difficult to fathom, but he felt it was true to the place and persisted with it. One day, legend has it, he found his own otter – an orphan cub rescued from a holt after the mother had been shot. The story goes that after hours of trying and failing to feed it milk from the nib of his fountain pen, he put it alongside his cat who was already nursing a kitten. Miraculously the cat allowed the otter cub to suckle. Later he fed it on fish, dog biscuits and vegetables. As it grew, it followed him everywhere, coming when he called and accompanying him on his walks. One evening, however, disaster struck. During a walk, the tame otter was accidentally caught by the paw in a rabbit trap. Williamson was able to release the panic-stricken animal, but he was badly bitten and the otter fled, terrified, having had three toes severed by the metal teeth of the trap.

Devastated, Williamson reportedly searched for his cub for weeks and months, following otter tracks along the rivers Torridge and Taw from the estuary on the North Devon coast to their sources on Dartmoor. He travelled vast distances and scoured every inch of territory he named 'The Land of the Two Rivers'. The story (and it may well be just that – a story) of this journey has a mythological quality. Whether it was part real or part imagined, it changed Williamson's life for ever. As he slept out, waking drenched with dew under the stars, he imagined his nomadic otter moving far and wide through the landscape. The memory of the otter and his unrelenting search scorched themselves into him. The whole obsessive episode reads like a sort of shamanic initiation. Scouring the land he became feral, and tracking like a shaman, he must have imagined himself into the animal's skin. During the process, Williamson dreamed up an otter that he named Tarka.

Williamson felt that his own storytelling had a strong element of truth about it, as if wild nature were speaking through him. In an afterword to the novel he explains his vision: 'This hand may have held the pen, this mind may have shaped the narrative – but of that which arose out of my living I felt then, and feel today, I am but a trustee.' The animals in his books are shaped and moulded by a whole history of stories connected to the landscape. Celtic people, whose spiritual life was rooted in stories of transformation, drew their mythology instinctively from the land and its creatures, and Williamson felt himself part of this honourable heritage. He always maintained that his bond to what he saw as a greater past was primordial. With *Tarka* he seems to have tapped into a traditional vein of myth-making where the landscape is rewritten with a spellbinding imaginative spirit, a combination of the actual, what the poet Ted Hughes called 'the icy feeling of the moment of reality', and the ancestral energy of something much older.

'From the worm to the hawk, the stone to the mountain, the cloud to the ocean, the fish to the hare, there is a mysterious web of connection,' the poet Alyson Hallett says. 'We may or may not forget that we are inextricably linked to all of these things through the blood and the bones of our ancestors.'

In *Tarka*, Henry Williamson somehow captured this. He caught

85

the objective reality of otters and created a poetic mythology at the same time. His creature of transformation slipped easily between the elements of water and air, inviting in the human imagination. These days our vision is so blurred by a combination of history, mythology, literature, Disney and nature documentaries that our thoughts can be trapped between cliché and mystery.

High on the moor it's harder to find signs of otters. They like to spraint on a tussock, a prominent rock, a silt beach or the entrance to a bridge, but up here there are none of these features. As it leaves the bogs of Cranmere Pool, the East Dart River becomes a widening pathway in a wetland of reeds and glassy water. I walk along its edge as it noses through peat and wobbles over pebbles, and the only sound is wind in the grass and a tiny whispering of water. I'm struck, in this desolate place, with the idea that these sounds have not changed since the beginning of time. However much we might have cleared and managed the land, ringed and counted every bird, identified and categorised every creature, the voices of wind and water are untouchable.

Otters might find tiny stickleback in the shallow water here, or frogs, or the glassy young elvers that will one day grow and swell into eels. In winter otters depend on the small amphibians that seek refuge between the roots of the rushy grasses. Where the moor has been grazed, there are many grasses growing: bristle bent, sheep's fescue and mat grass, but also tormentil and chamomile, and rare plants like heath violet. Soon, barely perceptible banks rise from the muddled whiskers of tall molinia and soft rush, making their way between moss, rocks, peat, lichen and liverwort. To my eye there is still no sign of otters on this open moor, only the ghosts of them, moving under the wide sky, with its mist and scents of cotton grass and sheep-fleece.

Following the narrow flow down from the bleak terrain of the high moor, the land begins to appear shaped by the river and gradually opens into a wide, gentle valley. There are some low falls, and long-decayed signs of human activity. The crumbling remains of tinners' huts and confusing trenches, the peat passes, show that people lived and worked here. What sort of isolated life it must have been I cannot imagine. Lower down, the path joins an old leat, a stone waterway engineered a hundred years ago to carry water to feed the reservoirs and tin mines. Apart from these ghostly remains, you can pass a whole day up here without seeing a soul. In the long grass of a marsh I stop to peer at an island in the river; this looks like a perfect otter haunt. As I step out towards it, my foot slips and I fall as vertically as a lead weight into the bog, finishing thigh-deep, saved only by my other foot which lands on something solid. Once I have extricated my foot with my boot barely attached, I walk some way trying to banish thoughts of being swallowed by the peat, imagining it closing over the top of my head.

With one leg plastered in brown peat and soaked in cold, I press on. Some trees loom into sight and I begin to feel better; these beech-es indicate the first human settlement that the young East Dart meets. The line of ancient beeches graces one side of the valley, and finally, accompanied by the pleasant aroma of my peaty encounter, I reach the road at Postbridge. This tiny settlement has a handful of houses, a bridge and a post office, as its name suggests. The post office sells snacks and it feels like a metropolis after the wind-blown place I have been in. I scuttle in to get myself some restorative chocolate. Inside, the postmaster and a foreign customer are deep in misunderstanding, so don't seem to notice my dishevelled appearance. The customer, who is German, is asking for 'shilly'. The confusion continues as we discuss this further, before realising that he wants to buy some *jelly*. The particular flavours of jelly prove even more difficult, as neither the German nor the English words for the fruits concerned bear any resemblance to one another. I leave them to it and walk to the old granite bridge, the first the Dart will travel beneath. Having eaten my chocolate, I scramble down to the water to scan for signs of spraint. From my position, I can hear laughter as the German leaves the shop loaded up with his many flavours of jelly.

A car rumbles over the bridge, and three cows eye me curiously. Scrubby woods of hazel, alder and oak have rooted themselves on the banks here, and the water of the Dart becomes more powerful as it surges into the woods, joining with other water from the high moor. Its route pushes through fields that are walled and grazed by ponies, cattle and sheep. A writer friend who lives on this part of the moor tells me that she glimpsed an otter up here. I'm not surprised to hear her story; she knows her part of the moor intimately. If she encounters anything unusual she will often send me a message. But here the river is narrow, fast flowing and hidden. It's very unusual – in fact almost impossible – to see an otter in this type of landscape. She was lucky. One day, out walking, she tells me, she happened to be looking at the water. Our eyes meet in mutual comprehension as she talks. When the otter slipped by it looked like it was surfing the current to save energy. She describes it moving downstream in a spate that was brown as milky coffee; one moment it was in amongst the waves, rocks and ripples, mingling its chameleon colour and shape with the flow, the next it was gone. 'I've been back every day to see if I can see it again,' she says. I shake my head; we both know that the chances of this are slim.

Later, she sends me a poem about it:

> a ripple firms into a jink of fur
> undulates into a clay-coloured pelt
> purpose flows from whiskers
> to the tip of a straight-ruddered tail
> If you wait and watch for his return
> you'll hear the river's many voices utter his name.

In *Tarka* the river Taw spills northward off the moor with the same magical intensity as the southward Dart in this poem. Williamson's river Taw develops 'long claws of water' as it pours downward, and we are drawn into its changing moods as it merges into the human world: 'Stunted trees grew, amid rocks, and scree that in falling had smashed the trunks and torn out the roots of willows, thorns and hollies. It wandered away from the moor; a proper river, with bridges, brooks, islands and mills.' The river becomes a journey between the wild and domesticated worlds, as Williamson paints its textures of savagery and tameness. Like the Taw in North Devon, the southerly flowing Dart is a spate river; rising water from sudden rain on the top of the moor washes everything away with fierce regularity. This means that signs of otter are often hard to find.

Lower down there are more human signs but none of otter; the strange striations of ancient field systems mark the slopes, and abandoned hut circles that have long since tumbled down are covered in moss and bracken. In other parts of the moor, rows of stones mark out where our Neolithic ancestors worshipped. Worn Celtic crosses wilt sideways. Now only a few areas are drained, farmed and walled with granite. Water is siphoned away in leats and conifers are grown in modest plantations between scattered outcrops of boulders and granite tors.

In a wooded natural cleft in the moor I find Dartmeet, where the peaty moorland waters of the East and West Dart join forces and become the Dart proper. The two rivers mingle and deepen to create a mighty torrent that twists its way through steep valleys studded with ancient oak. The river is beautiful here and distinctly ottery. Ferns, roots and boulders along the water's edge form nooks and crannies where an otter could hide during daylight hours. Plunging falls froth into deep-green salmon pools. Otters might forage here to catch the exhausted salmon at spawning time, rounding corners to find the sleepy fish and have an easy meal. I'm told that there are fewer big fish to be seen now, and otters must be finding other foods. In one cold winter, when many animals perished in the snows, deer fur and even cattle remains could be found in otter spraint around here. They must have been forced to feed on the winter's casualties. The adaptability of an animal will always determine its future, and the otter must be versatile, making it resilient to climatic changes. If food is scarce, they may cross the watershed looking for new feeding grounds. Ted Hughes describes an otter's coast-to-coast migration in his poem 'An Otter' where he imagines the wandering animal crossing from sea to sea in three nights. Years before this was written, Henry Williamson had imagined his otters crossing the moor in the same way, although I think it would have to be a very hungry otter that would travel all the way across the lonely marshes of Cranmere.

Hughes remarked that when he read *Tarka* as a boy it inspired him so deeply that he read it repeatedly: 'It entered into me and gave shape and words to my world,' Hughes recalled, and talked of how the language in *Tarka* put his life 'under an enchantment that lasted for years'. Hughes chose to move to the same part of North Devon as Williamson, and the two writers became great friends. Hughes's otter, with its 'Underwater eyes, an eel's oil of water body, … crying to the old shape of a starlit land', finishes with a fate as terrible as Tarka's: 'Yanked above hounds, reverts to nothing at all / to this long pelt over the back of a chair.'

The resonances between Williamson's tale and Hughes's poem are audible in the lines. In a generous memorial tribute after Williamson's death, Hughes said of *Tarka* that it was 'something of a holy book, a soul book, written with the life-blood of an unusual poet'. On Hughes's own death, his family arranged for a memorial stone of

Dartmoor granite to be placed equidistant from the sources of the rivers Teign, Dart, Taw and East Okement. Among the rugged gorges of North Dartmoor, it is deep in Tarka country. Its exact site remained a mystery for some time, hidden close to the source of the river Taw, amongst the marshes of the high moor.

Finding food is the biggest issue for an otter, and these days it is more likely that this will bring it into close contact, and even conflict, with us. The otter lives on the water creatures that appear through the seasons of the year; tench come upriver to spawn in high summer, and brown trout and salmon in summer and autumn. In winter it becomes harder, and the otter will have to forage further or for much longer. We know that otter populations are expanding, and overlapping with urban areas, and they can be attracted to our garden ponds. Most otters will be drawn to a free meal and will visit unfenced ponds, or even commercial fisheries, which can make them unpopular. Owners who live close to a river might be able to guess at who has been 'disappearing' their fish. Otters are used to grazing on mouthfuls of tiny stickleback and bullhead, but slow-moving food is best because it requires less energy to catch. A copious and lethargic ornamental carp is too tempting to pass by. It is also too large for one otter to eat. The owner of the half-devoured carp suffers an expensive and aggravating loss, with only some chunks eaten out of the fish.

The challenges for the otter are immense: it has to be large enough to maintain its body temperature and muscular enough to have the strength to feed itself. It has to be resilient and versatile enough to cope with sudden fluctuations in food sources, pollution incidents and other environmental changes such as floods and the encroachments of human activity. While it is awake, it must be constantly on the move. It may be big, but it still lives on a knife-edge.

And in spite of its bulk, the otter remains out of sight. A few miles downstream from Dartmeet, I find no sign at all. As I follow the spectacular gorge which takes the river to Newbridge, I realise that again the path has vanished, this time into some of the wildest, most impassable woods I have ever seen. I make my own way, following the river, clambering over rocks and through tangled brush. Bright holly and rowan berries light up the woods. A roe deer startles away and bounds up a dangerously steep slope, its white rump vanishing skyward. A buzzard wheels overhead, hurling its cry down the valley. I find a wide, flat rock to eat my picnic on and gaze down at huge fish moving upstream. I can't tell whether they are salmon or sea trout, but they are sleek and graceful as they move through the amber water.

November. I begin a search of the river's edge in the thickets of Hembury Woods. Here water is the colour of stewed tea. It syrups around boulders and slows into deep pools where the surface is bright with rust-coloured reflections of clouds, bare branches and withered oak leaves. This is the very edge of the moor, just before the river enters the human world at Buckfast, where the water is bordered with an abbey, several small factories and a waterside pub. I can feel the presence of otter, but never see one. For several days I go to the same part of the river as it flows through the woods and splits over a rocky beach to pass a small wooded island. Finally I find a sign! One particular otter – I assume it is only one – is using the tops of prominent rocks to mark its nightly patrol. The rocks it chooses are like megaliths; it is making no secret of its presence to other otters. It must have chosen these marking stations so that the signs are not washed away. I come at dawn and at dusk, and position myself carefully, where the otter will not see me. I have a full view of the otter stones, and should be able to see it clambering up. Each day I find new spraint, but never see an otter.

Meanwhile, other things come into focus. On my frequent visits to these stretches of the Dart, a heron forms itself out of a dead branch; a peregrine stoops out of nowhere and crashes into the ground after some unwary prey; a stoat lollops lightly along the riverbank; a trail of wood ants strays from an intricate city of pine needles.

At Buckfastleigh, a small market town, I visit a pub which is perched right over the river. I'm following a tip-off – a friend has told me there are otters in a cave in the cliff face at the water's edge. I knew nothing about the cave but, having asked permission, slip through the back gate of the pub and climb down to the water. The entrance to the cave is sandy and washed clean by the water. Inside, the roof height diminishes rapidly and disappears into fathoms of dripping darkness. I crawl a little way in and quickly begin to feel the pressing weight of claustrophobia. I light my torch and sweep it around the dank floor. Straight away I find tracks; an otter has been here. There are the prints of an adult otter and some slightly smaller ones, which mean cubs.

It's late November; the texture of the landscape is withering. My search should become easier because there is less cover to hide the otters on the river. Now they have to feed more because it's cold, and the otters will certainly be feeding more frequently if there are hungry cubs.

I've walked upriver from home, towards the otter cave, monitoring where the water flows wide, slow and shallow. The path is set away from the bank; it edges along a meadow which is sagging with end-of-season exhaustion. In the early morning, spiderwebs are veils weighed down by water. The night's rain has transformed parts of the field into a water meadow. Finches flit about, gathering last seeds from the frail tips of the meadow grasses, preparing themselves for the winter. I move to the bank to avoid wading through the floods and find that my new route, a narrow path meandering on the edge of the river, is exactly one otter wide. At the cave, there are no more signs. Perhaps they have got wise to me. The family has moved, but in which direction?

The clocks have gone back and darkness is drawing in. In this cold, it is easy to forget that our climate is changing. By the river a birch tree clings to a few of its uppermost leaves. Is the tree drawing its energy back from the leaves by increments, holding on to the topmost leaves till last? I walk quietly in my otter track and scan the water's edge for visible entry and exit points. There are no signs of spraint.

I place myself on a large oak root and look up at the trees. Intricate and extravagant, their elephant-skin bark conceals a highway for water sucked up from fathoms of earth below. It translates earth and water into energy-giving sap, consumes CO_2 and gives barrels of oxygen back to the atmosphere. In an essay on trees, Victorian naturalist writer Richard Jefferies remarked that one always found more birds around the oak than around any other tree. The oak's wrinkled trunk provides a home for hundreds of species of invertebrates, and a rich food source for a whole chain of birds and mammals. Its branches and leaves flutter and shiver with life. For us humans, oaks have been worshipped and used as meeting places and landmarks, and have given themselves for shelter, fuel and timber. They have nourished us with their beauty; we have recreated their forms inside cathedrals, barns and the hulls of ships.

I watch a brown-and-gold hornet come out of its nest under the bark. It tastes the air, flexing its mandibles, tempted by unseasonal sunshine to warm its armour. In this tree I can see three hundred years of growth; it is an intimate civilisation clothed in ivy, pennywort, lichen, fern and moss. All this from just one acorn.

The last days of November, and the land is going to sleep. I go a mile inland, up to the ridge of the watershed near my house. I want to know what happens to the water on its journey through the farmland that stretches down to the otter's habitat. Soon I am at the highest point for miles; Dartmoor's crags are visible on the horizon, and the valley of the river Dart rolls south through hobbit-shire roundness, meandering extravagantly toward the sea. The rocky green lane I'm following rises from near my home up through farmland and between the fields. An old drove road that heads in a roundabout, up-and-down way towards the city of Exeter, it is too sinuous to be Roman and too steep to tarmac or navigate by car. A few generations ago, this must have been all there was if you wanted to go to market, or to the next village. These ancient and well-worn ways, known as holloways, were made for human feet and horses' hooves; in winter they pour with water and in summer burst with a thick tangle of life. Only a few people use them now.

At the top of the hill I can see more of the crags of Dartmoor and south toward the sea. Here at this watershed, a spring fills an ancient well which is green with moss and liverworts. The spring becomes a stream and trickles down the lane, eroding banks that have grown over time into a canyon twenty feet deep or more in places. When it rains, a small river takes hold of this cleft in the landscape, deepening it by increments; more springs appear and all the water joins together and heads downward into a brook, which feeds into the river Dart.

It's quiet here and feels very far from all the roads. The green arable fields are grazed by sheep and cows; the once-creamy wool of the flock, the muddy hides of the cattle are all stained red with Devon clay. Whole fields are given over to winter beet and turnips.

Wind whips along the ridge. I count four skylarks twittering as if splitting notes and remaking them. A wren moves like a small brown mouse through ivy and black-knuckled ash twigs. Lower down the valley, between reeds and old flag iris, the stream ribbons into a bog. I catch sight of a slim fox loping back to the wood, its tail drooping behind like a lanky bottlebrush.

I can see that these fields are farmed intensively; tons of pesticides and fertilisers must go into this soil. The grazing fields are seeded with one species of rye grass; rain washes away nutrients from the ploughed fields, and each year they need more and more to maintain their fertility. Pesticides, nutrients, fertilisers, all have to go somewhere, eventually leaching downward in runnels and soakaways, toward the stream and the river.

Not far down the valley, in a tiny cluster of houses, I can see the tall tower of a church and the chimneys of my daughter's primary school. All around the village I can see so many colours, so many shades of green and grey, and all of them bleeding into winter. Many of the children go to school in cars now it's cold. We have decided to leave the car at home from now on, and we walk.

Left unused for a week or two, the car battery quickly goes flat and I don't feel inclined to fix it. Each morning we wrap up warm and walk amongst hedges of spider tripwires and frosted twigs. We notice somebody's stray white guinea pig in a field, count discarded banana skins and drink bottles, and step over all the broken bits from cars: wing mirrors, hubcaps, bumper parts, and things people have dropped as they drive. A used baby's nappy and some fast-food packaging lie next to a line of badger prints in the mud. Always in the same place, the badger tracks lead from a small wedge of woodland, under the steel gates, across the tarmac road and through a field on the other side to a distant sliver of trees. We peer at these paw marks and mull over the size of each badger. One day, we are halted in our tracks by the russet body of a fox. It must have been hit by a car in the night, but still looks beautiful, lying full length on the grey pavement. Its coat is glossy and mottled with the colours of glowing coals and ash. Its eyes are beginning to glaze. My daughter pulls at my hand; she doesn't want to look. As we leave I notice a bright trickle running from its nose.

As sleepy classmates are driven past us in cars, we arrive in the playground alert, pink-cheeked and full of the wild. The poet Rilke wrote: 'there is no part of the world that is not looking at you. You must change your life.' Other mums cheer us on as we walk; the teachers smile approvingly, but nobody joins us in our small rebellion.

December. The year is at its lowest ebb. Frost lingers in the fields along the river valley all day. Trees and plants have retracted their sap into the earth for protection. I cycle a little further upriver than usual, until I can see the crumpled tors of Dartmoor. The horizon is white. Grasses and twigs are starred with frost. It is winter, but there is still some of each season in the hedge. I can see primrose leaves and pennywort. A red campion is flowering.

I try to warm up my knuckles and walk down to the water, its icy smell in my nose, thinking about giving up my car completely. I'm able to get about without it, and more aware now that even a short drive disengages me from the land.

Low sunlight bathes the world into a frosted graveyard. My feet sink into paper-crisp layers of frozen leaves. I wonder why we bother with the rest of the day when there is this dawn. The earth is a crackling sculpture lit with ice; spiderwebs are like spun sugar, my breath liquifies against the sky in drifts of white. Grass seeds are petrified in frozen fronds; twigs are sparkling with cold, each and every fallen leaf and blade is laced with intricate crystals.

My dog accompanies me. He isn't thinking of the global economy, or what to have for supper, he's just happy to be outside, in a world of enlivening cold and multi-dimensional odours. Watching the sweep of his snout as he quarters the field, I can almost sense the web of trails he is following. And for once I can easily see the tracks: badger here, molehills created a few days ago; a fresh, sharp fox-scent in this direction; a pair of deer coming down to the water.

I find mink footprints at the waterside. Many people, spying a dark-brown mustelid galloping along a riverbank, mistake mink for otter. But otters are twice the size and can be ten times the weight. Otters have broad, blunt muzzles as opposed to the mink's highly pointed head. The mink is darker in colour than the otter, and has amazingly bushy fur, which makes it extremely buoyant when it swims, its whole length showing above the water, including its short brush-like tail. In comparison, the otter's long, smooth body stays mostly beneath the surface; often only the head can be seen, with an occasional glimpse of the back or the tapering tip of the rudder. On the occasions I have seen mink, it always seems to be more confident around humans than the otter, and will often continue with its business and not run off. I once saw a mink pursuing a rat along the riverbank. As I watched, the bedraggled rat scampered through the shallows, and the mink, a short distance behind, was relentlessly running it down. I had the impression that the pursuit had been going on for some while, and the mink, fixed on its prey, was not going to give up or hide just because a human was in the way.

Otters do not want to be seen. If disturbed, especially if they are surprised, they will disappear instantly. Mink are less shy, and their signs differ – mink tracks have thin, splayed toes and are far smaller than otter tracks, at about half the size. As it is difficult to see either animal, tracks and spraint may well be all we have to rely on. The mink's spraint are smoother, far more pungent, and twist and taper to a fine point, where the otter's often look more like small, crumbly chunks of partly digested fish or amphibian bones.

American mink also differ from otters in their breeding habits. They can have up to nine kits per litter, and the young mature and disperse within weeks, so they can colonise the otter's territory rapidly. They take far more birds and fish than otters, predating relentlessly upon ground-nesting birds, and will even take chickens and valuable specimen fish. Coots are in danger, as they nest beside the water, and the unlucky water vole, which burrows into soft riverbanks, is especially vulnerable.

A trio of mergansers floats by as if set on glass. The light catches on the crests and plumage of these slim diving-ducks. The males are astoundingly graceful with crested bottle-green heads and marbled black-and-white wings. They sit low in the water and their red beaks dip-dip the surface of the water as they pass.

A single crow shouts a claim on the morning. A cow bellows in return. There is no sign of otter. The water billows and swirls with oily blues and rust-reflections, like an abstract painting. I look at it so long that the opposite bank begins to flow in the opposite direction, making me dizzy. I notice I am surrounded by birds; a pair of robins are bickering in some intense battle, a wren flits deftly through twigs, calling *chink chink chink* at my presence. A party of tiny pink-breasted long-tailed tits fidget and whistle. I wonder what they can possibly find to eat. But later, as the sun lifts a little higher, everything comes into sharp focus. Groups of tiny midges dust the air, and I notice a dozen frost-capped molehills, turning up all sorts of microscopic morsels in beautiful thawing mounds. In one handful of perfectly milled earth, there could be hundreds of species: earthworms, mites, fifty to a hundred species of insects and invertebrates, nematodes, hundreds of species of fungi, and thousands of species of bacteria. Nature has developed a system that never needs to put extra fertility into the soil. It's all here. Twenty years ago the newly ploughed field nearby would attract hordes of rooks, gulls and raptors; now in places very few birds follow the plough. Just over the fence, the soil is eroded and exhausted. Here there is literally nothing for birds to eat. The effect of all the ploughing has been like taking the skin off a living creature.

Here by the river, none of this is apparent. The meadow is surrounded by woodland, and molehills rise like a small, anarchic civilisation – a counter-culture that ignores all the rules we impose. The presence of moles must be a good sign. Google 'moles', however, and hundreds of commercial sites devoted to the eradication of this little underdog of the mammal world will appear. I have never seen a live mole; they rarely come above ground. If they do, it is usually a young adult looking for new territory. The young will dig their way out of the natal burrow, burst out into the air and scuttle quickly away to establish a fresh territory. Being above ground spells danger for any mole. Sometimes, their stiff, velvety corpses can be found mauled by a dog, or 'exterminated' by humans, their little front teeth gritted against the light.

January. Dawn. I am watching a barn owl quarter the field. Snow covers everything. There is nothing to eat, and it continues searching in the daylight hours, desperate to find something to nourish its thin frame. A barn owl is thirty-five centimetres tall, with a wingspan of about a hundred centimetres, and weighs only 275–400 grams. That's less than half a bag of sugar. Hold one, and you are amazed at its lightness, the white-gold feathers, stealth-soft.

Humans have the ability to shape-shift, to adapt, to be creative with a problem; in this cold, these creatures do not. Barn owls are some of the most flexible and widespread birds, but faced with the whims of winter, they will feed or die. The barn owl lands on a fence post, its face turns, head swivelling, black eyes searching, like a famine child. It rises again on moth wings, fluttering toward the woods. There is nothing to eat. The voles are all gone deep into the earth; everything is retracted into itself.

In February I travel to what has been known for the past twenty years or so as 'Tarka country'. This area of North Devon stretches from North Dartmoor to the Bristol Channel, covering the wide and varied water-meanderings of the nomadic and fictional Tarka. It takes in the territory of his two rivers, the Torridge and the Taw, which eventually flow into a confluence and out into the Bristol Channel just beyond the market towns of Bideford and Barnstaple.

Henry Williamson may have resented the fact that so much attention has been focused on *Tarka* as opposed to his other books. He wrote many subsequent novels, including a momentous series of fifteen tomes entitled *A Chronicle of Ancient Sunlight*, which contain a brilliant sequence about the First World War; but it was *Tarka* that touched the public imagination the most. Now the railway from Exeter up to the coastal towns of North Devon is called the Tarka Line. Close-by, a cycle and footpath runs for a distance along most of the river Torridge – one of the two rivers that feature in the book – and this has been named the Tarka Trail. Williamson died in 1977 before much of this was set up, and it is impossible to imagine what he would have made of it all.

There's been a good snowfall, but miraculously the trains are still running. I take my bike on the Tarka Line from Exeter. This tiny, two-coach branch line is single track in places, and we chug thirty-nine miles across Devon, all the way to the north coast. I read the stations as they pass: Yeoford, Copplestone, Lapford, King's Nympton, Umberleigh. The names sound like a map of sturdy old England.

The journey meanders through country smothered in white. There don't seem to be many towns along the way, and hardly any of the villages are visible either. Instead, I get enticing views of the river Taw, and more snow, tinged reddish in places from the clay fields churned by grazing Devon cattle. After an hour the journey finishes at Barnstaple. I unload my bike, pull on my hat and gloves, and set off to find the cycle trail.

Snow can be useful to an otter tracker. If it has not thawed, or snowed again at dawn, the previous night's tracks to and fro will be visible. On the other hand, otters' spraint may be covered up. I stop for a late lunch in Bideford, and recharge again at Beam Weir with some sweet, steaming tea from my thermos. A sign reminds visitors of the fact that the fictional Tarka was born near here. I sneak off the trail and slide down to the river as it gushes over the weir. Out of sight of the trail, I find five-toed tracks into and out of the river, just below the weir. Tarka's descendants are still living here, in the exact spot where Williamson imagined his story. In the soft snow, some of the webbed toes and the longer hind feet are clearly distinguishable.

North Devon seems quieter, more exposed and less cosy than my patch in the South Hams. It is still winter here. Low hills roll to the coast, and the fields are dotted with beefy ginger cattle steaming around their hay feeders. Serious about surviving the cold, they have grown thick, rough hides of curly fur. Not far away, on the coast, tall cliffs are battered and weathered into gritty beaches by the brown churn and foam of the Bristol Channel. Further north, across the water, are the mountains of South Wales.

I cycle further along the Tarka Trail, far away from the coastal resort of Bideford. I've brought a copy of *Tarka* with me to read by the Torridge, and dip into it with freezing fingers. Williamson must have walked this river a million times, and I can almost hear his voice in my ear as, one by one, each of my senses is thawed: 'Time flowed with the sunlight of the still green place. The summer drake-flies, whose wings were as the most delicate transparent leaves, hatched from their cases on the water and danced over the shadowed surface.' The eye-widening observations are written with a kind of sensual reverence: 'Scarlet and blue and emerald dragonflies caught them with rustle and click of bright whirring wings.' Again and again, Williamson teases out our sympathies for the smallest detail of nature. British writer and mountaineer Jim Perrin, in his brilliant essay on nature writing entitled 'Holy Things', suggests that Williamson was unusual because he was ahead of his time: 'bodying forth his own views in brilliantly evocative, simple images, his social criticisms so slyly elliptical and attractive that, having read them, they become a part of our habit of thought'.

Williamson's writing certainly changed the way I thought. Even now, reading it quietens my mind's clamour, as the meandering calligraphy of the water and its inhabitants are drawn into sharp relief.

Williamson is alert to obsessively tiny detail: reeds iced like filaments in winter, a water beetle bending the surface of the water with its feet, gnats dancing in an elaborate evening display. And the river becomes an element of transformation, metamorphosing into an animal, where a pool 'quickened smoothly into paws of water, with star-streaming claws'.

The story moves us into a dimension where wild nature is meshed beside and around us, and Williamson, attempting to pass on the spiritual understanding that he had of the natural world, paints beauty and savagery, heroism and destruction. Humans were as much a part of it all as the heron that swallows the eel or the starving otter that attacks a swan in the middle of winter. The destructive powers of man are sharply drawn during the hunting scenes, which contain all the horror of battle and of killing. Here is an unspoken outcry for the victims of violence. Williamson may not have felt able to speak out against brutality, but the message is implicit in the lines. Years later, the American poet and visionary nature writer Gary Snyder called for children to be educated more explicitly about all this, saying that it was not enough just to know about nature, emphasising that we need to feel it all the way through. And on this point, Williamson was a perfectionist. If his hands were tied, his words broke free, and came directly from the heart: 'Each word,' he said to Ted Hughes about *Tarka*, 'was chipped from the breast-bone.'

Later, I turn back downriver and scan broad mudflats, rich with hundreds of wading birds. As the tidal zones open out, the estuary is fed by silt from the confluence of the two rivers and all their tributaries. Wide skies blow into changing clouds and salty air. I stop at bridges, find otter spraint, photograph it. The mudflats are littered with prints. The sea is close. Now designated a UNESCO 'Biosphere', the whole of Tarka country is protected for the unique richness and diversity of its wildlife. This area, from moor to sea, includes the entire catchment of the Taw and the Torridge rivers, from the sources to the estuary. It even includes a landmark in the sea – a new marine reserve which reaches all the way to a tiny outcrop of rock in the Bristol Channel – the puffin sanctuary of Lundy Island.

Perhaps the idea behind this Tarka merchandising isn't such a bad one. The cycle and walking trails unite human activity and the wild with a degree of harmony. Travelling along the route, the human body is toned and worked, while the environment remains relatively unpolluted. What better emblem for the whole thing than the nomadic otter, 'wandering as water', Tarka?

As I pedal along the changing river, my skin becomes permeable to the world's freeze. Williamson's words are like a soundtrack in my head, bringing the wild alive in this cold. The subterranean stories are like a tissue woven in the whitened banks and icy ditches around the river, in the frozen fields and meadows, in the woods and the dark stream as it runs down to the sea. Claws of winter grass line the banks, mosses work their magic, birds flit about searching for a stray seed or insect to eat.

108

Close to the encroachment of busy coastal settlements there is space for otters to survive. A stone's throw from the road verges there are banks and water hidden below; here the only traffic is bird-sound, wheels on the well-tended bike track and the quiet breath of cyclists. At Beam Weir I stop and talk to a birdwatcher who is filling up an impressive array of feeders with nuts and seeds. He tells me quietly that three otter cubs were seen here this year. Hundreds of cyclists pass this landmark each week, with no idea of the covert mammal activity nearby.

In spite of human impact, the otter never died out here as it did in many parts of England. The rivers of Tarka country have provided watery corridors that were free from disturbance and protected the otter from threats. The North Devon farmland that borders these rivers is private, so it offered exactly what otters need: undisturbed countryside which keeps these animals from contact with us. Many of the watercourses that flow here are constantly cleansed, springing from the wild uplands of Dartmoor. When all else fails, the otters can still retreat along the waterways into the shielding moor, where the untainted landscape enfolds them.

At home, my river moves like a quiet animal between the banks. Its energy is low. If you sit by the river at this time, in unseasonal sun, you can feel its tiredness. Upstream from the town, a fading notice says 'no fishing', and explains to those who didn't already know that fish stocks are dwindling.

The river will not give up any signs of otter. Perhaps they have gone in search of food, or a new mate? Like the deer, they have melted away. Increasingly perturbed, I go three or four miles downriver to a tidal stretch with wide mudflats and oak-tangled banks. There might be more feeding down here for otter, with the daily flood of the tide and its nourishing silt. Alder catkins dangle brown tails from bare twigs. A rosehip store hangs desiccated by the winter. A pipit surveys me from a damp woodpile. On the grass, some fungus that looks like oyster mushrooms lifts slimy cups to the sky.

A raven circles, hurling a guttural *kronk, kronk, kronk* over the fields. The banks of the creek are scattered with seaweed where the brown tide has seeped out. Dried-up chestnuts strew the path like tiny porcupines. A little gull mews across the mud-water, flaps up to tussle with a mate. The clay in the fields here is a rich red, and sticky on my boots. Cattle have made paths to escape from the mire, but everything is saturated and clogged with mud. The rotting carcass of a sheep lies in a mess of wool and ribcage. In the woods, somebody has nailed up a handwritten notice about a local man who died. The paper is crinkling at the edges, and at the base of the trunk somebody has laid some flowers. The notice says: *Life and death are as inevitable as the seasons of the year*.

A hen pheasant creeps along the hedge, blending her ash-mottled plumage into the old leaves. The hedges are backlit with lime-coloured hazel catkins. Cloud mounds up against blue, light strays across water. There won't be many days like this. The light is acute and glassy, the year poised for the end of winter, tipping into something edged with sunlight.

I search for otter at the river's edges. A skeletal hawthorn, drenched in lichen, leans over the waterline. Beetles eat away at the trunk, life and death mixed everywhere. Part of the bank has fallen away and the red earth smells like clay and incense. I find more catkins and flick their little tails, watching the green puffs of pollen pour into the air. The tide slips out over mudflats which dazzle with light. Finally I find a slide – a fur-smoothed chute in the bank where an otter shape has slipped down to the water. Its eyes may be watching from beneath the overhanging roots of the oaks that line the opposite bank, but for now the river guards its secrets.

Marsh

Beast lopes down riverbank
runs on webbed feet
puts her seal on the mud
but entering the water
she enters her name

Paul Hyland, 'Otter World'

In Devon I have glimpsed only one actual otter. It would be nice to see a whole otter for more than a few fleeting seconds, and I've found out that there's a watery place in England where you are more likely to find one than anywhere else. It's not far from my home, in the neighbouring county of Somerset. Here, apparently, there are large swathes of otter country I knew nothing about. I'll have to go deep into the wetlands, and some serious tracking will be needed. Any new territory has to be carefully read. Field signs (the tracks, spraint and any other evidence the otter leaves) will have to be located first, but I know that even this has its challenges. And then there will be the sitting and the waiting.

In my search over the last months, there have been so few signs; sometimes the ground remembers the otter and sometimes it doesn't. There may be a few tracks by the river, as if the creature manifested itself, touched the ground with three feet and then vanished. There might be a subtly defined pathway, trailing away from a tangle of grass on a riverbank, or, barely perceptible on the air, a fragile scent. With-

115

in a few hours even otter spraint can be desiccated by the wind, odours vanish and the tissue of hints evaporates. But with each tiny clue, I fall deeper into the spell.

Television footage makes it seem easy; the otter pops up in the water at close range, or potters its way into the viewfinder of a night-vision camera. But tracking the otter is more complex than this filmic view implies. You must spend a good amount of time finding nothing; then and only then, perhaps an ambiguous sign will turn up. I might notice a cryptic trail through the moisture in the grass in the morning; long grass might be bent over the wrong way; there might be a shadow where something left or rejoined the water beside a bridge. If I am exceptionally observant I might find a sign in the undergrowth, a tangle shaped into a tunnel, and if I peer close, I might notice it is rounded to fit an otter. There might be a few scratches, sand turned into a mound where scent marks are still damp, or a twist of grass with spraint close-by. On the other hand, often, there is no sign at all. I have to depend on my developing understanding, my nose, and a kind of intuition that has begun to develop these last few obsessive months, of where otters might be.

On the riverbank, the five-toed mark of the otter's webbed feet can be hard to identify. The claws are blunt from turning over rocks and pebbles on riverbeds, so they do not appear clearly, as in a dog's print. The toe-prints simply taper to a point, finished by an imperceptible claw. Also, because the otter walks in a pigeon-toed fashion, leaning on the edges of its feet, often only four toes show up, the fifth one barely marking the ground, which can be confusing. Otter toe-prints

116

are tear-shaped, and together they form the width of a spaniel's print. However, an otter's toe-prints are much more splayed than a dog's, and are more supple and malleable, so its prints can vary. The toes are arranged in an irregular semicircle around the front of the foot. Sometimes there may be a ridge at the front of the front pad, but the webs, the one feature that distinguishes the otter print from that of many other mammals, are almost always invisible.

The otter's hind feet are larger and more elongated than the forefeet, and from a distance they can look like flippers. They are designed for propelling the owner like a frogman through the water. When it dives, the otter does not always use its forelegs as much as the hind legs, and the forepaws can be folded to its chest for more streamlined dynamics. When it is swimming underwater the otter uses its whole body in a rippling motion like that of a dolphin. Subtleties invisible to us, like the speed and strength of the water, the density of the salt in the sea, or the waves and currents, all affect how the otter swims.

Back on land, the otter's tracks are only ever visible if left in soft mud or sand, and are most often found around bridges, or on the beaches beneath them, and on exposed or silty parts of riverbanks. You can find them if it hasn't rained, or if the water level has remained the same. Where I live in south-west England, rain often washes away any evidence. In tidal areas, it can be even harder. Licked at by the river's edge, prints fade like an unfinished sentence. In *The Wind in the Willows*, Kenneth Grahame's otter regularly vanishes mid-conversation and does not return for days. Although anthropomorphic, this well-observed eccentricity is more realistic than we might think. Grahame was a naturalist, and he knew that the otter is not concerned with making itself understood, or helping us to find it. It only wants to communicate in its own language and only with other otters. If the riverbank is rock or shale there is no visible sign at all, but sometimes a narrow otter track may be worn to the earth or the grass smoothed from where the otter comes out of the water. These pathways are often used to cut off a bend in a river or to cross a field. The otter has to save all its energy for the demands of fishing in cold water, and if there is a shortcut, the otter will make use of it. These straight paths can travel a mile or even two from rivers, but it is not possible to mistake them for human paths as they are too subtle. When I began tracking, I used a strange little path beside my local river for a while before I realised that it was an otter short cut.

The only truly reliable way to find otter, often the only sign that remains for any length of time, is the spraint. In a one-mile-long stretch of my river, I found six different spraint sites, but without access to DNA testing I could not easily tell if it was one otter or more. Some research has been done into the differing smell of individual otters' spraint, but the results aren't yet published. I thought at first that this particular researcher must be extremely dedicated and have an unbelievable sense of smell. Now I've found out that although this may be true, in fact the work is aided by the technology of mass spectrometry. This machinery breaks down the chemical elements of a substance and separates the important features. I love research scientists. Choose any aspect of the natural world, be it size variation in the individual toenails of woodlice in the Ashdown Forest, or damage to the eyesight of crustaceans caused by algal blooms in the Bristol Channel, and there will be researcher looking into it. Where I live, a woman recently won a prize for research into the homing instincts of snails. It gives me hope.

The otter usually spraints after feeding or where it comes out of the river; these signs are punctuation marks in its journey. They can be found beneath a bridge, or on a ledge, a mound, a prominent rock or large tree root – in short, all along its own well-worn pathway, denoting which way it went and what it fed upon. The otter's fast metabolism means that it digests its food very quickly, so there is always a clue as to what it has been eating. There could sometimes be tiny rodent bones, but more commonly the spraint is made up of parts of the fish skeleton itself: very fine hair-like rib bones, tiny vertebrae, gill cases, fins, and minute teeth and scales. Sometimes there are feathers, or even the fur of small mammals. All this will vary with the time of year. In winter and early spring there are fewer fish, and amphibians are a popular prey. Frogs that have been eaten are recognisable as small clumps of tiny broken bones. In summer there could be more fish, as well as what could be the contents of the fishes' stomachs. Crumble spraint in your palm and you may notice indigestible remains of what a fish might have been eating. These might be beetle cases or tiny larvae.

Regular visits to the otter's favourite tussock, rock or outcrop will yield telling results. James Williams, in his exhaustive otter-spotting manual *The Otter Among Us*, suggests creating your own 'otter loo'.

He writes that he has made a very successful one at his riverside home in Somerset, allowing him to monitor the otters in his patch. You place a flat rock above the waterline where you know (or suspect) the otter usually comes out of the water. If the otter passes through, it should leave its visiting sign. This sounds like such a good idea that I plan to set one up in my next hunting ground.

I also plan to visit Williams, as I think he may be a kindred spirit. If his book is anything to go by, he has huge otter-knowledge and lives very close to the Somerset Levels, an area that appears to be an otter hotspot. My journey there takes shape in my mind. Asking a local expert is always useful, and having read his impressively detailed book, and the information in his regular bulletin, the Somerset Otter Group's 'Newslotter', I think we will get on. He will know where to point me to start looking for otters.

I spread a map over the floor and stare at the contourless grid of ditches. *Somerset flats*? *Plains*? What was it called? Only a few of my friends have heard about the Somerset Levels, and they are birdwatchers. It's a strange and haunting place. Not much of it is over eight metres above sea level.

Although it is just over an hour's train journey from where I live, the landscape is described in a language of foreign words. Small outcrops called 'mumps', 'bumps' and 'burtles' rise out of the mire. 'Rhynes' (pronounced 'rheens') divide the fields. Neolithic people built a sparse network of narrow oak walkways over the saturated clay and peat so that they could travel through it with dry feet. These paths were barely one person wide and supported precariously by stakes driven into the mud. One would have had to walk with vigilance. They sound like the otter paths I have found. Narrow and practical, they draw you intricately into the landscape, threading through it with economy and lightness.

The Somerset Levels. I need a car to explore this network of tiny lanes, and driving there, I realise that this is land where water evens itself out constantly. Humans have spent generations tipping the balance. Years of draining and reclaiming and it is still waterlogged. Low-lying, flat and striated with water that seeps imperceptibly in all directions, it is the largest wet meadow system in Britain. I like this. A land that has been interfered with, divided and channelled, but is resolutely leaking back to its natural state.

From the road, it does not look like a glamorous location. It is a two-dimensional topography of grey cloud, grey road, grey water. The M5 shoots through without pausing. If you lose concentration for a minute you could blink and miss it. A giant wicker structure – a man with gargantuan pectorals and no hands – fashioned out of locally produced willow is positioned as if the man is running alongside the motorway, as if to say *just keep going*. For many people, there wouldn't seem much point in leaving the seductive trance of the road just for fen and marshland, but I can feel my excitement building. All this flatness suddenly seems exotic. You can see for miles. The level fields are striped and criss-crossed with the rhynes, ditches and canals that have been used as 'wet' hedges for hundreds of years. It's a maze of water, and perfect otter country.

I've realised that one of the reasons I follow the otter, apart from being fixated with its wary secretiveness, is for its superb taste in places to live: the seductive wild places, mountain streams or granite cliffs, moorland, estuaries. Exploring this new area is not the heart-swelling experience I'm used to. It's flat and featureless, an altered, broken landscape where humans have drained and shifted the water table, and built dodgy, soggy pasture where before there was none. The rigid surfaces of roads and obstructing buildings that have sprouted here in recent history do not make any sense to the otter. It is drawn to feeding grounds where there are fish. If these are in the stream on the other side of a road, and it is unable to get through a sluice, the otter will go straight across, regardless of traffic, which has produced inevitable casualties. The Environment and Highways Agencies have put in special reflectors all around the roads to dissuade otters from ambling into the beam of headlights at night. They have dug up roads and put in underground crossing tunnels or platforms wide enough for an inquisitive otter to trot over. They have installed fine-mesh anti-otter fencing between the streams and the road. But sometimes the otter will choose not to travel securely. It might go around the fence, through a gap, make use of an open gate, or avoid the tunnel because it smells of a bigger otter. When it gets to the other side of the road it may be trapped by the fencing that is supposed to protect it and have to wend its way along the tarmac.

The tourist guide to the area suggests activities such as a visit to the Peat Moors Centre, bird watching or fishing. The Centre sometimes has otter information, I hear, but is hard to track down, tucked away between villages in a mesh of little lanes near Shapwick. When I get there I find that it closed months ago, but inside I can see the remains of beautifully reconstructed Iron Age round houses, one collapsing appropriately back into the peat.

The old peat mines that are dotted around have been flooded, and now contain reed-bed havens for aquatic wildlife and birds. Close-by there are several useful reserves with hides that look out over these large areas of water. With interconnecting pools of water everywhere, this is one of the soggiest places I have found to look for otters. The watery veins and capillaries provide flexible highways that permit them to slip quietly amongst us. Here it seems they are unstoppable: they have been seen in towns as well as outside them, seeping through our gardens, ponds, pools and workplaces, following water with a freedom that ignores roads, walls and culverts.

The otter was here long before us, and in spite of persecution and our toxic activities, reports indicate that it has been returning to its old territory in most counties in England. The Somerset Levels are no exception. This tamed grid of slow-moving waterways is just as attractive to otters as any of the places we consider to be wild. 'The wild' is no longer 'out there' in the way we might think; those places that were empty, untainted and free from humans are no longer so. As we have expanded and colonised, the wild has become knitted around us, in a living, breathing mesh. The otter is truly among us. The Levels have been broken up and mined for peat, but they still provide something akin to what otters need. The fields are in a constant cycle of water; they are alternately drained and reclaimed, their water continuously, quietly changing. These changes otters can often deal with; after all, what river was ever the same on different days? Otters are built for change. They can adapt, mingling with the subtle flow and wash of this fluid land.

I drive along feeling strange about using my car again. It's so flat that next time I think I might come on a bike. To the south and east I can see sodden peat fields and marsh and to the west the rounded Quantocks. Further to the north, the rumpled line of the Mendips; Cheddar Gorge is in there somewhere, with its underlying formations and fathoms of limestone. Beneath it all, water makes its subterranean ways through rock, percolating, eroding ventricles and chambers inside the secret rhythm and drip of the earth. I feel a sort of vertigo when I think of it: the depth and the inscrutable swarming dark.

The underlying rock is like a water-carrying bowl. Sometimes it fills and the water tips out. There are still seasonal floods, and have been for thousands of years. Evidence shows that Mesolithic people came here in the summer, when high winter water levels dropped. Now the area is tamed, we use it all year round, and rework the ground, digging ditches in the saturated layers of organic matter to divert and reduce floods. But the water always returns. In the natural course of things, the wildlife simply adapts, fits in or disappears, swallowed into the wet mouth of the bog. The 'vowel meadow', as Seamus Heaney calls the ancient layering of bog, has a million-year-old meaning and language all its own, and he imagines the wet centre of its depth as bottomless.

Out of the car, with the sour scent of peat in my nose and the wind chasing its tail through the reeds, I study the hieroglyphics that an

124

otter has left. This flat land seems to tilt you always towards water; the opaque strangeness of peat, the warp of the tidal mud, the wobbling graphs of water levels and ditches. I slide carefully down a trail of prints towards the rhyne. The peat is spongy and can bounce back any prints left by an otter. I'm lucky to find these and, further on in the grass, a small day-old spraint.

Later, back home, I phone James Williams and introduce myself. Here is a person, I can hear instantly, who is a mine of information. He answers my stream of questions with so much detail that I give up trying to write it all down and we arrange to meet instead. When I arrive at his house the following week, the first thing I notice is a weathervane in the shape of an otter fixed poignantly on the roof. Williams has dedicated a large part of his life to mapping the weather of the otter's fortunes. At the back door, there is a row of wellingtons and a cluster of tall walking-poles, which I later find out are essential kit for the otter professional. 'Just come in,' James calls from somewhere inside, 'we don't use the front door.'

We sit in the kitchen and get straight down to business. James opens files over the table, spreading detailed printouts of the ongoing work he and the Somerset Otter Group have been doing. Outside the window, nuthatches and blue tits flit back and forth. 'Look at the statistics,' he says. I look at dizzying tallies, results and graphs with their sharp rises and dips over the years. The detail is immense. Somerset, on the eastern margin of the West Country, is an 'otter frontier' county. From here, as the population expanded and re-established itself, otters could have spread north and east from their stronghold in the South West. Although otter presence is relatively strong in Somerset, James reminds me, the population is still vulnerable and many otters are found dead.

'Almost all of the otter deaths that we know about are caused by accidents on roads,' he points out. 'But why should they have been on the road? What caused them to leave the water? And the ones that die quietly in a ditch, we'll never know.' James is licensed to collect otter corpses (you need a licence, due to the otter's protected status) and, in conjunction with the Environment Agency, regularly sends dead otters to Cardiff University, where a dedicated research team studies each body to record details of the animal's state of health when it died. Elizabeth, James's wife, must no longer be surprised when venturing to the freezer to take out the Sunday joint, only to discover something amorphous and less than appetising.

James has been surveying an area of over 4,000 square kilometres since 1969. He has produced detailed annual reports on the findings, and written two very brilliant and helpful books about this elusive animal. It is clear from the reports and what James says that there have been large fluctuations in the Somerset otter population, and there may be a similar pattern for the rest of the UK. Many rivers in Somerset went for a long period without any otters, and there have also been sudden and dramatic declines, with no obvious cause.

The first national otter survey, which took place in 1977 in response to the vast decline in otter numbers, revealed a disastrous picture. Nationally, less than 6 per cent of sites yielded positive results. No signs of otters were found in many counties for twenty years.

James's graphs show that for almost ten years between 1978 and 1988 otters had almost completely abandoned the Somerset Levels; estimated numbers were as low as two or three. This calamitous pattern could still be repeated in the future, James warns. Flooding, agricultural and industrial pollution, loss of habitat, persecution, fighting and road-kill still affect numbers.

The worst culprits are cars. Over the last ten years, James tells me, the Somerset Otter Group has logged an average of 31.7 road deaths per year, out of a varying population now estimated to be about 65 otters for the county. Cub births help re-establish numbers (James estimated 22 cub births in 2009), but this is precarious too. Otters are quite slow to reproduce, and it may be that they have to travel large distances and cross roads and river catchments to find one another to breed. There are many more questions than answers, but amazingly, in James's 2008 Newslotter, he was able to estimate that with approximately 65 otters in the county there must be about one otter per 57-square-km section that was looked at.

So otters are still scarce, and they seem even more precarious when you look at their breeding behaviour. They only have one litter of cubs each year, and there may only be two cubs in each litter. Statistics show that about 42 per cent of cubs die in the first few months of life. Many are mysteriously abandoned. Unlike other mustelids, female otters reach sexual maturity at around two years old, and look after their cubs for a year, so the capacity for colonising areas is much slower than that of many other mammals. Otters are also highly vulnerable to man-made disasters such as toxic waste and accidental spills of any kind. Many otter corpses are found to be infected with toxoplas-mosis. This disease originates from cats when faeces from their litter are flushed into the water system. In 2002 research showed that this action had killed up to 13 per cent of Californian sea otters. When toxic human effluent arrived in the sea from the western United States similar research showed that otters were also vulnerable to some hu-man diseases. When humans competed with the sea otters for prey, and the otters were forced to eat shore birds, many of them died from the parasites that these birds carry.

Where water rules, the otter is like a pulse within. Like litmus pa-per, it gives us clues as to how the ecosystem is doing. It reacts to slight changes in the climate of the water and to our behaviour. The new national otter survey may show a slowly recovering otter popula-tion but, James worries, this could give a false sense of security. There is so much that we can't predict, he says. Humans are inventing new toxins all the time, and as these build up in the ecosystem, we do not know how they will affect our wildlife or its habitat.

We are outdoors now, wearing our wellies, walking down the drive towards the stream where James is going to show me his handmade otter loo. He is still telling me about toxins. The more recent ones to be discovered in the watercourses are polybrominated diphenyl ethers, or PBDEs; these are organobromine compounds used as flame retardants. PBDEs are used in an amazingly wide array of products: building materials, furnishings, carpets, clothing, cars, plastics, poly-urethane foams, resins, sealants, packaging – in short, in most things. Research has shown that these chemicals are toxic to the environment; they build up in tissues of living organisms and cause behavioural and hormonal changes that have affected all animals that have been tested. High levels are found in human breast milk. 'In the meantime,' James says, 'we're all pretty flameproof.' I giggle. 'But water voles have disappeared,' James goes on, 'so have eels.' He watches me from the corner of his eye to make sure I'm still listening.

We look at the fresh otter spraint on the rock James had placed in a spot that he knew otters used. 'My otter loo,' he says with some pride. I notice the addition of a sand trap, where James has made a low wall and filled the space with sand to catch any prints that might stray across in the night. He waves his stick towards the tangled roots of an ash tree not far off. 'See under there? Otter holt.' I peer into the dark bundle of roots, but we don't approach. 'It doesn't look like much, but you need a licence to go near it.'

There are other reasons for not approaching that begin to dawn on me. Our footprints may make a noise, upset a subtle balance in the riverbed, or just encroach on the otter's private space. James knows this, but I hadn't seen it so clearly before – how humans crash through the invisible boundaries that animals know to respect.

We go back to the house and James gets his car out. On the back window is a sticker which says 'I Could Kill an Otter.' There is mud inside the car, and some pheasant feathers, which I choose to overlook. We travel to Burrow Mump, an outcrop close to the A361 where you can look out over the wide flatness of the Levels. Burrow Mump rises up inexplicably high out of horizontal fields, and nearby there is a confluence of two rivers. These are the Parrett and the Tone, which used to meander sinuously through the ancient flood plains on their way to the sea, a few miles away. Now they are straightened and canalised, like many of the rivers on the Levels, which is not so good for otters.

In winter the fields between these rivers are deliberately flooded to maintain the delicate wetland ecosystem; ubiquitous sluices are closed to trap water, and water birds flock in their thousands to wade and feed. I speed up to follow James, who is already striding up to the top of the Mump to show me the view. A ruined church dominates the summit and we look down over wide, icy flats. The ground is a map of frozen pools dotted with tufts of willow. The ice is peopled with flocks of wigeon, lapwing, teal and other waders. There must be at least fifty swans. I spot the white rump of a solitary roe deer feeding beside a clump of withies.

The February ground is all mirrors and thawing ice; an otter would find plenty to eat amongst these flocks of waterfowl. James points to the place where the latest otter casualty was found. Attracted to the good feeding, one was killed on the road close to the water sluice which controls the stream and the water level of the nearby fields. Unable to get to the winter hunting grounds through the water-filled tunnel, the otter climbed up, crossed the road and was hit by a passing car. I watch the speed with which traffic hurtles along this straight stretch. The otter didn't stand a chance.

Back at ground level we put on fluorescent jackets and walk to the spot where three otters have been hit in two months. Now one more otter has gone, but we find fresh spraint at several places nearby. Beside a rhyne, James points to a collection of cracked swan mussels lying in a jumbled heap. 'This is an otter's midden,' he tells me, picking up a shell. Sharp teeth marks pattern the shell where it has been prised open for the meal inside. *An otter's teeth have been here*, I think, slipping a shell into my pocket.

Otters may have died here, but more will quickly move into any vacated territory. Somewhere nearby, curled in a holt of willow roots, brambles and wet grass, another otter is no doubt sleeping off its feast. At night it will emerge, a shape as fluid as water, foraging, extending its range, moving over fields, downstream, into and out of the town, crossing culverts and roads in its search for food.

As otters are encountered more and more frequently in our urban environment, they are caught on CCTV, or seen scampering back to leats in the early morning; they have even been spotted attempting to catch ornamental carp from garden ponds. In Exeter a dog otter was discovered on CCTV, visiting the Wildlife Trust headquarters at the town mill. He had been using the old mill leat as a passageway, and climbing through the mill wheel. It seems, from the camera footage, that this otter only came on Saturday nights, when the Wildlife Trust employees weren't there. In the few seconds' footage, he moves confidently; this is his territory and he looks like he knows he will not be disturbed. This suggests to me that he lives very close; he knows about our human routines, and overhears our comings and goings. During the day he could be hiding out, closer than we imagine, in a nook in a wall, or on an old building site, while we continue our noisy daylight lives, unaware of his presence.

There used to be at least ten watermills along this old riverside industrial stretch in Exeter. Otters would have been back and forth, finding the same ways through for generations: they would have learnt to blend in with the ever-increasing urban activity, slinking around the Roman walls, slipping through the docks, swimming subtly past the flour, paper, powder and wool mills. Now modern streets with bijou flats and shops replace old mill names that echo days long gone: Cricklepit, Powhays, Surridge and Exwick. When the river Exe was diverted by town planners, the otter must simply have found its way and carried on.

In Taunton, the county town of Somerset, otters have left signs near the doctor's surgery, on the golf course and at the public library. One, a bitch otter, was hit by a car outside the County Hall. She had two cubs; one was found dead and one was rescued but died later. 'This,' James says, 'is why my first book is called *The Otter Among Us*. It was here first, long, long before us, and it carries on in spite of us.' Tracks can be seen by the town's bridges, and traces of what has been eaten are visible from the river walk. We might hear an otter calling in the dark, but tucked safely inside our houses we never see it and our ears would probably not distinguish the strangeness of its whistle from the dusk calls of birds.

I leave James and drive out of town, to where the otter's landscape spreads through fields riddled with waterways, and the land dips towards the coastal towns and into fluid flatness beyond Bridgwater and finally to the edge of the Bristol Channel. Here the otter forages in the slippery inter-tidal zone. It's not ground and not water, shining with a slow seeping that is almost impossible for humans to negotiate. Strange wooden sledges called 'mud horses' still navigate it, sliding fishermen out to collect their catch and roll in the nets safely. Sometimes an otter may take a grey mullet from a net and have a feast. This does not make them popular.

The land is eroded and smoothed, washed by water finding its way to join with more water. The otter runs through all of it. It crosses the flats light-footed, slides quietly into water smooth as a toboggan, front feet folded to its chest, back feet paddling. In her poem 'The Riverman', Elizabeth Bishop sees the secretive movement of animals like the otter and the silty water of estuaries as blurred, unarguably clever, and enchanted with a strange 'more-than-human' life: the liminal place where the river inhales salt from the sea and 'breathes it out again' is described in a heady song imbued with shamanic vision, where 'all is sweetness, in the deep enchanted silt'.

Far from the towns, a lattice of water feeds the peat moors of the Levels. Some of this area has been named the Avalon Marsh, as if everything is charged with something we have forgotten or lost, as if water and forgetfulness have mingled. For most of us in Britain, water goes obediently through pipes, gets filtered in strange, churning digestive systems and comes back out of a tap tasting of chemicals. There's no more fetching and carrying of water, no religious homage. Our knowledge and experience have gradually marginalised marshes and swamps, and at the same time they purify water and replenish the land without any of our technology and energy-burning regalia. The marsh is an ingenious filtration system. It stores carbon. It is living memory.

A reading of the layered chapters in the peat reveals the story. After the last period of glaciation, sea water rose and this land became a salt marsh. Then as the sea slunk back, it reverted to rich swamp, building layers of goodness supporting a million species of organism. Later it became fenland, then woodland grew here and there, and eventually the ground, carpeted in a miniature rainforest of sphagnum moss, laid down its archives in layer upon layer upon layer of rich, dark earth. Humans came, and discovered the locked up energy; where the peat was mined, and the ancient sunlight released, water seeped back to heal, creating a new patchwork of pools. Rush and reed beds grew in a new meshing of life. Elizabeth Bishop described the healing powers and wisdom of water in her talismanic poem 'The Riverman', observing that as it came from the heart of the earth, the cure for every human need, including the remedy for all diseases, could be found there. The water here is subtly playing out its role in the geography of this place, and as Bishop said, we just have to find how to look and to see it more clearly.

The otter is one part of the river that we need. Like an emissary from the swamp, it has something to teach us. It indicates that all is well or not well. Unseen yet close-by, like a spirit level, it tipped and diminished, but it came back. We have reached a point where our relationship with this animal is crucial. I want to follow it, but it crosses to where I can't go. It is a mercurial teacher; it draws me in, then disappears.

Much of my time looking is about losing the path. I feel like a novice Buddhist monk. Wading through this animal's territory, I can't even see my own feet. I make mistakes. The ground sucks at one of my wellington boots, and I topple over, soaking my exposed sock. This horizontal land looks firm but does not always hold. It plays tricks. In some shady spots the lack of reflections makes it look bottomless. Untrustworthy layers of sphagnum moss float on water. I want to feel the solid bottom but I am not sure where to tread.

I can't look at the marsh without the stories of its dark side creeping in. At night here it's as black as a bag, and you can't see or feel your way out. In the fog it feels as if the earth wants to eat you. Our ancestors used to throw votive offerings and trinkets into the mire to avoid being devoured. In *Beowulf*, Grendel comes out of the swamp to drag people off and feast on them. Bogs did and do still swallow people. The Grendel stories translate wetland into a dark, mapless world: 'it is not far from here,' the story suggests, inviting us to glance over our shoulders, 'nor is it a pleasant place.' The memory of the devil-ridden mire, the unconquered swamp, has always been close-by. On the other hand, the American writer Aldo Leopold, in his *Marshland Elegy*, admired marshy landscape so much that he claimed he would have liked to *be* a muskrat. Henry David Thoreau loved to stand up to his neck in a swamp. He said that when he was dead they would find bog oak written on his heart; and Seamus Heaney sanctifies the ottery bog as part of his national identity. He describes its fathomless texture as saturated with another sort of language 'meaning soft,/the fall of windless rain'. Does the shape of the watery landscape affect the way we feel and see? These writers at least seem to have been consciously nourished by wetlands.

Dorothy Wordsworth wrote about the watery landscape when she lived in Somerset. Just south of the hills where she was living, water was being mapped and colonised by engineering; rivers were being tamed and exploited, canals built, waterways diverted and made navigable. The dominant discourse was about control of the rivers, ignoring most of the wild species that inhabited them. Beginning her Alfoxden journal in January 1798, Dorothy writes: 'The green paths down the hill-sides are channels for streams. The young wheat is streaked by silver lines of water running between the ridges.' She is highly sensitive to water in her writing, and notes it flowing freely in all the nooks and crannies of the landscape. There is not a diary entry which does not mention it. Going to fetch water from her well, cooking with it, washing and cleaning with it, working on the everyday domestic arrangements for herself and her brother William, she notices every drop. When she writes about their daily walks, she notices how the moon affects the tides, but also observes the attributes of a dew-encrusted snowdrop as it gradually begins to break into bloom. She knows which plants are thriving in the moist hedgerows as the season turns, and even marks when the sap in each tree pushes it into bud. Thoreau, in his essay 'Walking', claimed that he did not feel right unless he spent at least four hours a day wandering and observing. The Wordsworths seem to have felt exactly the same. Free from the distractions of air travel, electronic gadgetry and digital technology, all these writers had a connection to and intimacy with the water and wetlands near their homes.

James telephones to tell me more otters have been killed on the Burrowbridge road. We talk regularly now. 'It may be road-kill,' he says, 'but what we do not always know is *why* these animals should have been crossing the road. Was it lame from fighting, weakened by hunger or sickness from parasites? Was its way blocked by something, or was it simply avoiding another otter's territory? We can't afford to stop asking questions.'

I go back to the spot on the Levels where the otters have been killed. This time, I decide to take the train and cycle to the tiny hamlet of Burrowbridge. At this pace, the human history of the area – the devotion, battles, victories, defeats – is palpable. At nearby Athelney, also known as the Isle of Athelney because it used to be an isolated island in the marsh, there is a monument to Alfred the Great. He famously took refuge here, in the year 878, and went on to defeat the invading Danes. The famous Alfred Jewel was unearthed here in 1693; Alfred had several of these intricate and beautiful ornaments made as special gifts to his loyal bishops. This place gave up the only survivor of these precious relics. The peat swamp has its own pages of memories, stories and ghosts, most of which it guards jealously.

I started out here in bright daylight, but the still afternoon drifts into evening and it's bringing a swirl of fog. I notice faintly colder air rising, and some darkening ripples in the pools around me, and I have an urge to turn back and head in the direction of the city and its lights.

This quiet maze of water and willow-studded ditches has been a swampy and impassable marsh since medieval times. The river Tone was diverted centuries ago by the monks in the nearby monastery to alter the boundary of their lands. The otters, it appears, favour the old course of the river. They leave their signs along a watercourse which is just visible as a rhyne with unusual kinks and meanders; it has not been flowing for at least six hundred years. It might be that otters prefer this route because it is more secretive, or that there is some memory residing in the sodden ground, some vestige in the wet earth that only an otter can read.

Where I live, the curvaceous clay hills of Devon draw me into feeling intensely embodied and engaged with the world; here, the flat lands of Somerset do something different. I'm an outsider, like a scholar studying a strange manuscript. I've been reading and hearing about the landscape, but now I begin to feel how everything has been flooded for the winter; strange sounds rise from flocks of waterbirds, like voices babbling from pools. A couple of swans barrel in, low and heavy as Hercules planes; they land, spectacularly, in the thawing water.

I search for signs of otters and spend some time puzzling over what can have happened to the most recent one to have come to grief. The low sun gets in my eyes. A merlin floats past and lands on a dead tree not far off. Cars rip through, making use of the long straight stretch of road to maximise their speed. Is there anything more to be done to avoid these accidents? If we can't or won't get out of our cars, perhaps we can at least switch off the music and slow down.

Later, back home, my Google News Alert sends me a tiny but disturbing item from a local Somerset paper. Unusual numbers of otters have been found dead in Somerset. The incidence of orphaned cubs has increased. It turns out that during autopsies, a new parasite, a fluke named *Pseudamphistomum truncatum*, has been found in the gall bladders of three dead otters. Otter corpses have been found, emaciated or prematurely dead. Autopsies reveal the fluke embedded in the otters' vital organs. James tells me that it is thought that the fluke has spread from escaped ornamental fish.

A research scientist from Cardiff University sends me a paper with pictures of damaged otter livers that have been recovered from corpses. Of the 237 dead otters examined, 90 per cent of them had the fluke. The livers, opened up, reveal a delta of damage and disease. On my screen a map of Britain appears, red-spotted with deaths spreading thickly over the South West and out as far as East Anglia, with some reaching all the way to Northumberland.

The wetlands are a vital organ; they are a sump of biodiversity, they transform dead matter into new life, they are vast carbon stores protected by a living skin and are the cleansing drains of the land. In the case of this exotic fluke, they have been infected by urban society's actions. Thoreau said that the swamps are the wildest and richest gardens that we have; he thought of wetland as the purest and most alive environment of all. The 'fenny labyrinth' he talks of in his writing perfectly describes the regenerating wetlands right here on the Levels. In contrast, he saw the city and people as bringers of disease. His vision was prophetic.

James tells me the best places to see live otters on the Levels: the protected nature reserves of Westhay, Shapwick Heath or Ham Wall. 'You might not see any,' he warns helpfully, 'it depends what they're doing.' He wants to come with me, but this I have to do alone. Two people tracking otters means twice the noise, twice the scent, and worst of all, the irresistible temptation to chat, which James and I always seem to do at length. On the phone with James, time disappears; the outside world blurs and our conversation meanders obsessively around otter anecdotes and the intricacies of otter habits and doings. It would be impossible to sit silently with him; inside a bird hide we would risk losing the carefully folded demeanour that otter-watching requires. The sides of the hide would billow with gales of laughter, the otters would flee and in the end we would clomp to the pub in muddy wellies, to continue the conversation over lunch. No, to achieve anything useful this trip will have to be solo.

My journey by bicycle on the tiny roads to Shapwick is almost entirely free of cars, and the earth feels calm and spacious. Moving around like this I can recognise and identify birds and plants, and cycle past roe deer and rabbits without disturbing them.

These reserves are very much alive; pockets of recovering landscape, like a healing wasteland or a rebuilt battle zone, they have been tended, studied and cared for. In many places nature has made its own way back. The wide, wintry horizons of the Levels seem suddenly full of possibilities. I cycle through trees, marshes and flooded fields. Shapwick Heath is a major wetland reserve, forming a large part of the Avalon Marshes. When more of the ice cap melts, this area could be permanently flooded again. American nature writer Gretel Ehrlich's book on climate change, *The Future of Ice*, explains: 'Water seeking water, that's what we're seeing. It's nothing new. Like everything and everyone, the river is filled with longings, no matter that they come to nothing.'

Two thousand years ago, from the top of Glastonbury Tor, this area would have been a flat sea of silver reflecting the changing moods of the sky, punctuated only here and there by a tor or a tiny human settlement. Eel and fish reigned supreme. Waterbirds wandered like herds on the African plains. People never travelled very far. Perhaps now, in the twenty-first century, nature is working slowly toward this again. Just a bit more CO_2. A few more cracks in the ice. The bubble of the spirit level could be wobbling towards its centre.

Dominated by water and its cycles, the reserve boasts a rich set of habitats, including herb-rich grassland, ferny-wet woodland, fen, scrub and ditches populated by a plethora of aquatic plants and invertebrates. Drier parts of the site are grazed by hardy goats or shaggy Red Ruby Devon cattle. These stocky auburn cattle are grazing by the side of a ditch as I slow down to look. One of them raises its head and paddles toward me; it looks at me with big eyes and licks a nostril. Other heads turn. They cluster in front of me, blinking heavy lashes, clouds of vapour rising from their grizzled hides.

Otters have to eat more in the cold winter months because of the energy levels required to survive in low temperatures. Moreover, in winter there are fewer fish, and the otter will have to spend more time foraging. I am now much more likely to have a sighting in daylight as a lot of undergrowth and cover has died back in the winter months. Where I lock up the bike, the land is ribboned with water; the smell of bogs and ditches soaks the air, drawing the gaze close-to, focusing me on the detail of surfaces glossy with wet. The path lures me into the reserve; I know otters have been sighted here recently. My feet follow part of the web of lines that link ditches and drains into a lattice of hazel coppice and stunted oak. The ways seem to lead from everywhere to everywhere, and I choose a path toward a wide expanse of water that has been brought into being by centuries of rain. I pull up my hood as the clouds loosen and souse me with all they can muster. Then the wind picks up, bashing the reeds and bending them into a whistling chorus. I wonder whether this will spread my scent or dissipate any human sounds. Squelching toward the camouflage of a hide I narrowly miss stepping upon a frog shining as if varnished with water. As I peer down it pings away and becomes a wet leaf among other wet leaves. Soon after, a bright light rinses everything white. Thunder follows. I can hear water sluicing underneath the hut as I creak inside. In the dark I fumble to unsnib a catch, open the flap and peer out. A world slick with water and rainbows seeps in.

I settle onto the bench and release my binoculars from their case. In front of me is a slab of grey water bordered with tall reeds. A whiteness drifts up from the shallows and spreads into the wings of an egret. I sit for hours meditating on the water surface. Periodically everything blurs, then clears. The reeds move imperceptibly, oozing with continuous sibilance.

The water changes from moment to moment. It is grey, it is ruffled, it is polished pewter or a mirror holding the sky and bouncing light in every direction. I am mesmerised as it furs with the lightest shower of rain, ripples beneath coots or bends under the weight of a swan. Moorhens bicker at the edges of my vision, and mallard mischoreograph landings, skating over the water in threes and fours. Gadwall flock and feed. Mostly, nothing happens. But nothing is good. I drop into stillness. My mind empties. The rain on the roof is a thousand pattering fingers.

A ragged battalion of cormorants perch on the sagging skeleton of a drowned tree. Rain begins to come in through my window, polishing the sill into the surface of an infinity pool. My toes begin to go numb. I zip my coat tooth by tooth up to the neck, pour myself some tea and eat a biscuit by melting it on my tongue rather than crunching. Still no sign of otter. The logbook notes one feeding in front of the hide yesterday, and preening its fur on the bank two days before that. I rest my chin on my fist.

The water surface is zinc, and brighter than the sky. Between the two, thousands of starlings are stirring. It is nearly time for me to leave. The starlings begin their pouring flight over the reeds. They are a flickering brown stream. Some settle like extra leaves in the scrub, others continue on to some invisible gathering place in the fields. I leave the hide as the sun is beginning to dip and the clouds are tinged with lemon. Colour shifts to etch in some ochre, then gradually daubs everything with a watery wash of orange and salmon pink.

Emerging from the woods I catch the smoky display of the star-lings. They form and dissolve against the sunset. Their spectral swarm seems to contract, then burst wide open. Watching the News recently, my eye was distracted by the same formations in the background behind the newsreader. Starlings were taking over the shot. I don't know what the story was, but the starlings rewrote it that night. I wonder how many other viewers noticed and were distracted by the shapes made by the birds? Far more interesting than the story that was supposed to be news, it was a protest in starling, a wordless counter-current to what was being spoken.

I stand absolutely still. Behind me the fading sky is piled with mountainous cumulonimbus. The birds are gargantuan against the clouds for just a few moments before falling to earth. Like the under-water movement of the otter, they leave me longing for something that slips away. As the birds disappear, a badger bustles out of the trees and forages for beetles in the soft earth of the track. Its sett must be in one of the banks in the coppice woods, and I think about the frag-ments of history it burrows through nightly: Neolithic, Iron Age, Bronze Age, Roman. Peat has been dug here for centuries. In 1970 an old timber trackway was discovered by Ray Sweet, a peat-cutter. Dendrochronologists dated the tree rings in the oak planks to 6,000 years old. It is possibly the oldest road in the world, extending across the marsh between what used to be an island at Westhay and a ridge of higher ground at Shapwick. Thousand-of-years-old feet creeping across the marshes. Perhaps searching for beaver or otter. Evidence of a network of these tracks has since been found, some of them obviously laid on trackways that were older still. Hunter-gatherer ancestors would have crossed and recrossed here, carrying their young, moving their families, hunting deer, trading pelts.

If you stand on one of the promontories, you can imagine what the topography must have looked like – wide, silvery expanses of water, criss-crossed by these thin lines; raised wooden walkways linking outcrops and their people. Otter, beaver, elk and wolves would have waded freely. I wonder about the library of memory held in the ground. It's a dizzying thought. In Bruce Chatwin's book *The Songlines*, he talks of when the Earth was a dark plain, and the ancestors created symphonic pathways which literally sang everything into creation. All we have left is a little archaeological evidence and small puzzle pieces. Chatwin says, 'an unsung land is a dead land: since if the songs are forgotten, the land itself will die'.

Listening to the quiet echo in the reeds, I can't help thinking of it shadowed with patterns and songlines. Moving like a fluid map, reeds whisper. Water trickles. A bird flickers through the air. It is a breathing web and neither the map nor my words can do it justice.

Night falls between the watery dykes. When my eyesight no longer serves me, I listen to the strange layers of song: the call and return of a pair of tawny owls, and later, the percussive high-pitched whistle that could be an otter, hidden in the dark. Something is rippling and reforming in puddles and pools. A thumbnail of crescent moon has risen. The poet Alice Oswald calls moonrise the 'hinge-moment' when different voices begin to speak and the moon hangs in mist and over water like a peeping eye. I find a bridge and wait for the whistle of the otter call. The landscape fades into shadows and greys, but the water still holds a memory of brightness. New sounds occur that I can't identify. The whirring flutter of something, a rasping cry, a twitch; rustling.

I am listening for something in particular. Otters have a range of calls; the '*huff*' seems to be when the animal is curious, surprised or slightly threatened, but also appears to be used as an aggressive response when it is annoyed. The high whistle is used for locating and summoning; cubs will use this sound repeatedly. There is also a high-pitched whickering, what Henry Williamson called a 'yicker', which is used defensively, when the otters become agitated or frightened, or when they are fighting.

Time blends into the dusk. I wait until I can no longer feel my feet. My senses begin to confuse and lose focus as my eyesight no longer serves. Then, like a soft clarion call, it is there. My heart misses a beat. Could it be? What bird would whistle at this time? There it is. It's the otter's whistle. I can feel my pulse racing. It's close. The whistle is repeated again and again, at intervals of a few seconds, in the same high, monotone call. It sounds urgent, insistent. There is no answer, only the trees rustling with crowded roosts and tangled foliage. My feet, half submerged in mulch, are glued to the ground. I wait and wait. I hardly breathe. Without the dazzle of a torch, the moon is all quivering reflections. I stare over a bridge at the flicker beneath.

Suddenly, a shape forms, leaving a rippled 'V' behind. A bulge in the water rises then disappears. I daren't move. The water is all the

146

colours of night; black peat with its archive of colour mixed over millennia; water, mint, algae and reeds make brown, mud and silt and sedges make grey, eels, frogs, stones; the down of bulrushes; all of it comes to black. How will my eyes ever see a moving shadow from the belly of the river, against this background? But there she is, a shapeless strangeness in the water, and a fluttering of discomfort in the reeds.

Another whistle, and the otter is joined by another. I cannot see them, but remain still, hardly daring to breathe, cursing the vapour of my own breath. A splash as the otters meet face to face in the water. I hear a yicker as they greet each other and their bodies break the surface. I think it's a mother and cub.

The faintest reflection of light catches their movement as they twirl together like rope. Otter cubs can sometimes stay with the parent until they are eighteen months old. At this point they are fully grown and have enough hunting skills to be independent. But the mother often has trouble getting rid of her young. They are clingy; they sometimes pester, follow and call to her for weeks.

A sequence of leaps and dives ensues, where the otters spar and bicker. The splashing is unselfconscious now, and unmistakable. One seems to have caught some prey and the other is begging for a piece. Otters do not like to share food. There is quiet while one chews and the other forages for itself. I can still only just make out their shapes. There is a flickering movement of jaws before they swallow and dive again. For a moment I think they have left, then they surface once more and I make out two long shapes, one just ahead of the other. They wend their way further down the waterway before insinuating themselves back into the dark. A breeze moves the reeds and I hear one more whistle, then nothing. I wait a little longer, until most of the light has gone. Still the water holds a faint glimmer of sky.

On my way back to the car park I'm light-headed with my otter sighting. Two otters together! I'm buzzing with excitement and can't wait to tell James. They might be otters that he knows about already, or perhaps not. Either way, this night-time encounter makes me think it will be worth coming back for more.

Just as I reach the road, a badger noses out of the undergrowth and sets my heart racing again. It bumbles away along the path in front of me. Striped nose fixed to scents in the grass, it follows an invisible trail laid down by generations of other badgers. I follow its ghost-greyness close behind until it pauses mid-mouthful, catches my scent, then bulldozes off into the brambles with a snort of disgust.

In the trees and reeds around me, hundreds of starlings are roosting. At dusk you catch them gathering on wires and branches, shifty, like Hitchcock's birds. They are still now, like reams of black beads. In the darkness my sensations are suddenly all askew. Not being able to see my own feet properly gives me a kind of vertigo. I'm not sure how much further it is back to the road. I can only function by being porous to what is around me. I try to bring the surroundings into my senses by a kind of osmosis. Animals must have the power to do this a hundred times better than me. This process of night-time perception reminds me of Steinbeck's image of how a human might gather delicate marine animals: 'so delicate that they are almost impossible to capture whole, for they break and tatter under the touch. You must let them ooze and crawl of their own will on to a knife blade and then lift them gently into your bottle of sea water.'

I let my mind go gentle, become a receptive vessel, allow sounds and shapes to 'ooze and crawl' on their own into my senses. Only then can I feel what is there and try to bring some of it into focus.

Shapwick Heath feels like a land that is recovering from vast human interference. Ecologists might say it's in transition, where ecological succession is still in a dynamic cycle. In the scars where people mined for peat, water has seeped in and evolved new ecosystems. Plants have colonised, along with amphibians, invertebrates, fish, eels, birds and mammals. I mention to an old friend that I am writing an 'otter's-eye-level' account of the area. She is a botanist, and listens bright-eyed, computing what I say. 'How far off the ground would that be?' she asks, moving to the bookshelf. I consider her question, watching a book entitled *Bryophytes* being opened up on the table. After half an hour, our eyes peeping at one another over a mound of books on botany, I agree to take her with me when I return. She wants to show me some ground-level detail.

A week later we set off in our wellies, armed with a botanical magnifying glass, to nearby Westhay Moor. Here I notice straight away that the earth has been gouged and dug up in various places as the rhynes have been cleared; large spoil heaps lie around, and there's a strong sulphurous stench about them. This interference seems brutal, but they are drainage ditches after all and need to be maintained.

The first thing anyone would notice is the unsightliness, and then the stink of decaying matter. My reaction is because it's not aesthetically pleasing to the eye. This is a nature reserve after all; surely it should be pretty? As we stand breathing in the smell of what seems like yet more devastation caused by humans, my friend tells me that this is simply the smell of land at an in-between stage of ecological succession; water has been overwhelmed by plants which are turning the area into a marsh, which eventually, if the water table remains low, will become dry land. Choices have been made as to which species are to be encouraged and nurtured, as a balance is sought. The transition between any of these phases may well not be pretty, but it is very good, I am told by my botanical expert, for the restoration of wildlife.

So we take our noses closer in, nostrils alert, and inhale the pungent low-level forest saturated with moisture; a colourful tangle of dark-red sundew, liverworts and mosses looks like some kind of primordial ooze, or the interconnected organs of a living creature. What she shows me is hidden amongst it all: one very tiny plant. The identification handbook says it is *Pallavicinia lyellii*, or Veilwort. Its leaves are like little spongy green tongues. We kneel down and water immediately soaks into both knees of my trousers. The sourness of the peat and wet grass gets right to the back of my throat. Growing subtly on the waterlogged earth between tall tussocks, the Veilwort is just a tiny, creeping, slimy fern-type of plant, but it's endangered and therefore precious.

My friend tells me the correct botanical names for the plant and all its parts. I hear 'Liverwort', 'thalloid', 'undulate' and 'papillae', but the rest go over my head. It seems a uniquely human trait, this urgent naming of things, and I start to question it, as if somehow naming is artificial. 'No,' my friend tells me. 'If we know what it is and what it needs, we can save it.'

It seems that an intricate configuration of needs makes this plant highly sensitive to any changes in the surroundings. It could vanish at any moment. It grows in ditches around sphagnum moss, and some-times even grows hidden beneath it. It likes the shade given by purple moor grass, soft rush and the old rush tussocks that survive here at Westhay.

All that fragility. We need to get down on our knees and pore over it to keep it safe. On the way back to the path, I find myself treading carefully, aware that each footprint left behind is slowly filling with water. Each new pool is a lake to the minute life that is there in the soil, each footfall creating a cliff, a quarry, stamping on one world, making another. We pass a pair of dog-walkers, and then a cluster of birdwatchers staring into the sky.

The next day I set out early. I've borrowed some neoprene waders, though I don't intend to fish. I want to look from water-level. I pack up the bike panniers before dawn and arrive as a thin line of light peers over the horizon. Following a tip-off, I wrap my feet in tinfoil for insulation. It can get very cold standing for long periods in waders in the water and apparently this is the thing to do. I slide into my neoprene membrane and waddle as stealthily as I can to my spot. Already birds are beginning their wake-up calls. Edging into opaque water, I am drawn into the rustling of reeds in the dark. I can feel cold clinging to my legs.

I stay for a long time, the kind of time that makes time seem like nonsense. Sometimes I feel unsteady, as if the silt under me is shifting. In their foil coating, my feet feel like Sunday roasts. Gradually the light comes up. Indigo becomes cobalt, which brightens to the kind of blue glow from a computer screen; the blue of a low gas flame burns for a while, then eventually fades to pale English-winter blue. Now the colour is lifted by hundreds of starlings flying out of their roosts. They flow above my upturned face like feathered arrows in clusters of ten, thirty, fifty, the reeds vibrating with their flight.

When the birds have gone, there is not silence; water and air can never be entirely silent. There is a constant breathing of reeds. I'm inside the forest of stalks and my feet, although protected by the foil and the wellingtons, are losing normal sensation. I am suddenly thinking about the hot sweet tea in my thermos. In the light of morning my stomach gives a huge gurgle of longing. I decide to give in, and in that moment, it happens. Something is moving on the water. It is among and between the swaying stalks, twenty yards away. I freeze back into my footprints and will my stomach to be quiet. A brown eye is floating on the surface towards me. It forms into a head, followed by a long body. I can see upturned nostrils, a half-submerged snout and two ears. The whiskers are webbed with water. I think it is unaware of my presence, but I cannot be sure. A tail tip leaves the water as it steers through the shallows. It bends into the depths, dips and comes up all teeth and whiskers, chewing what look like small eels. Even as it eats, there are no ripples.

It is coming closer, out of the reeds, towards me, almost close enough for me to touch. I have never been so near to a wild otter, and stuck thigh deep in its territory I begin to wonder, through my held breath, about the sharply uneven contours of needle teeth. Then I start to think weirdly: perhaps it is so close because it is a tame otter that has been released? Just as this thought dissolves, it looks up. Something about it stiffens. We are eye to wild eye; its face is armed with a startling array of walrus bristles. Its ears are larger than I expected, almost like a cat's, and its nostrils are visibly measuring my scent. There is nothing shy about this animal. I have got close enough to see five different sets of whiskers around its face and under its chin. In its eyes I can see shock at what on earth I am, and at what I could be doing in its hunting ground. The live current in both of us prickles. When I do not move it comes a little closer, *huffs*, then melts bodily into the water surface, leaving the shadow of a ripple and nothing else.

Back in the car park, hot sweet tea dribbles down my throat and thaws out my hands and feet. I grab my notebook, focus my eyes, and make a clumsy sketch of the otter as best I can; the picture is shaky, as if it should wobble or pour off the paper. I must be printed upon that otter's retina as it is upon mine. I wonder how long I will stay there, the untrustworthy outline of me, meaning nothing but all the contours of danger. I trap my picture between the pages, as if even this image is slippery. Then I set about peeling off my waterproofs. I lay out the foot-shaped foil for later use, put on a warm jumper and go to lean on the bridge.

The wet body of peat gives gently under me. Its crypt-smell speaks of a thousand years of history. The otter has been here through all this time; it was here before people came, before villages and towns and roads grew up. It flows through its water pathways, makes use of our canals, travelling unseen, moulding itself around us like a shadow. It has been among us since the beginning. As far as it is concerned, we have only ever been one of the earth's dangers; like Ted Hughes's otter, who

> *Wanders, cries;*
> *Gallops along land he no longer belongs to;*
> *Re-enters the water by melting.*

Source

She becomes water
becomes other,
becomes otter, fleet
in the pool and flow of river
the whirlpooling, spooling
unravelling river

Paul Hyland, 'Otter World'

I'm going west, to look for otters around the water sources high on the Cornish peninsula. This muscular arm of land pushes ruggedly out into the blue Atlantic. With its variety of craggy coves, clean freshwater streams and long sandy estuaries, it's perfect tracking territory. It's March, and energetic weather systems are doing their best to move as much rainfall as possible inland from the Atlantic. Heavy precipitation in this region means that streams in the wild granite uplands are frequently replenished and usually flowing with cheery enthusiasm. The geology of this far end of Britain is mineral-rich, and the aquifers were at one time plundered and polluted by the old tin- and copper-mining industry. Now many of the watercourses are pure and habitable once more, and the otter population is doing almost as well as it is in Devon.

Cornwall is known for being a 'stronghold' for otters, and it's thought that here, as on Dartmoor, their populations held on for longer than in other areas where they became scarce. But a stronghold? The word's deeper associations – of conflict and battles – have little to do with why otters have survived here. If the otter has any stronghold, it's simply water, and there are no fortifications for this. Otters are still vulnerable to human actions. At this moment in time I feel as if they couldn't be more vulnerable; on the radio I've heard discussions about a revival of mining in Cornwall. This was the industry that powered the local economy in the past, but the market collapsed and one by one the mines were closed. The mine closures left parts of the human population bereft, but at least the ecosystem was able slowly to recover. Recently, the worldwide price of tin and copper has rocketed and the idea of reopening disused mines has become interesting again. The copious mineral deposits in fissures in the granite all over western Cornwall are often below the water table, so what also needs to be extracted is water. In the past, at the height of the industry, more than 600 steam engines were being used to pump water out of the bedrock.

In this narrow peninsula the underground aquifers are never more than twelve miles away from the coast and a network of arteries flows continually out to sea. The water is affected by climate and human use, and both impact upon wildlife. It's a delicate balance. The slightest change can have repercussions, and the human population has increased, so that now more water than ever is being used.

Just before I leave, the phone rings. It's James, with news. There has been a pollution incident in Somerset. Gallons of something noxious have leaked into the water of a reserve where otters live. What will happen when the otters come through the surface of the slick and ingest the toxic cocktail of wood stain and fire retardant that will be clinging to their fur? Strangely, James tells me, it appears that only

158

one species of fish is floating belly-up in the poisonous concoction, and that turns out to be rudd. These first casualties are surface-feeding fish, so the pollution has perhaps not yet reached the layers in the ecosystem underneath. James found no otters, but beside the water where pollution had collected, something had been rootling at the turf, tearing it up, burrowing as if in a frenzied attempt to clean itself of some cloying substance.

I put my bike on the train, sit with my head against the glass and watch the early spring landscape. Rivers, trees and birds flick by. At Plymouth we rattle over the sturdy Royal Albert Bridge. This impressive portal between Devon and Cornwall was completed just before the death of its designer, Isambard Kingdom Brunel, in 1859. The beautiful wrought-iron bridge was quite an achievement. Its double lenticular design was one of an array of similar technological advances in the Victorian era, and for more than 150 years this bridge has been supporting railway traffic thirty metres above the waters of the Tamar Estuary. For all this time, the eyes of millions of travellers have been drawn in a wide sweep across the confluence of the rivers Tamar, Tavy, Lynher and Plym. During the final stages of its construction, Brunel's bridge attracted admiring crowds of thousands of onlookers, but legend has it that Cornish guests were prevented from attending the opening ceremony because their train broke down in Liskeard. There they remained, trapped over the border in Cornwall, several miles short of their destination.

I gaze down at the wide brown waters that divide Devon and Cornwall. On both sides of the bridge, the sheer volume of water is astounding. This landscape has been gouged and carved by water for millennia. Springs, streams and creeks dribble and pour downward, breathing in salt and exhaling fresh water in a pattern as complex as the internal workings of lungs. Once, humans found the shape made by the river useful as a territorial division, but although the most patriotic of the local tribespeople might disagree, it's no longer as necessary as a barrier. Its importance has been superseded by the coloured demarcations of maps and by our thirst for tarmac, travel and petrol. Below me, on both sides of the river, rubbing up next to the watery area of outstanding natural beauty, are commercial harbour developments, Europe's largest military port, international marinas and thriving boatyards.

Here is the chaotic meeting place of many elements. The familiar and the dangerous brush up together in this fluid and liminal place. Amongst the salt marshes and intertidal zones there isn't a real division between land and sea at all, but a meshing together of the human and the wild. The river's songlines and the tides sweep tons of silt and pebbles in and out, continually altering and fertilising something which we might think is fixed but is in fact dynamic, shifting and evolving.

The train teeters high above it all. The dysfunctional jumble of human living is everywhere, and all of it depends upon water from this river to survive. Dozens of toxins pour into what is left. In the past, zinc, copper, cadmium, iron, sulphuric acid and arsenic escaped as the Cornish tin mines penetrated the water table, and in 1992 a plugged mine burst and disgorged ten million gallons of contaminants into the environment.

I used to work down there. I taught in a local school and each day drove past the boatyards, the patches of railway land, the tangle of weeds, tracks, gardens, old roads and new roads, a naval base, nuclear submarines, outsized factories, towering cranes, dried-up reed beds, schools, chip shops, bakeries, moorings, pleasure boats and rows of terraced houses. Wrapped all around the edge and slithering into it is the bit I never looked at: the shining tidal mud of the estuary. Now, from the train in the morning light, I can see that it is whorled and patterned with tiny tributaries; everywhere they doodle and etch the fleshy sides of the river with patterned veins. The fluvial processes have transported and deposited particles in layers and textures of lithic memory; living organisms and floating sediment mingle and support one another in an endless cycle of accumulation and erosion. Plants, invertebrates, shellfish and crustaceans have adapted to the changing salinity of the water and the shifting ground here; some of them will be rare and highly specialised, forming a fragile membrane of life all over and through the mud.

Somebody has ploughed across and dug for fishing bait. The shiny surface of the tide-smoothed silt is pocked with an uneven path of puncture marks and excavations. A flock of gulls floats over the water. I can just make out the curves and pathways of the estuary as it dwindles away upriver, its tributaries, wooded creeks and hidden folds all seeping seaward.

Freed from the concentration required for piloting a car, my vision moves with the train, attracted to the fluid shapes caused by water in the landscape. Some of them are hidden from view in valley bottoms, some are winding alongside the track; streams suddenly seem to be everywhere. I can see the grace and wildness of them as they move, like some other tribe; sometimes soft, sometimes sinewy, concerned with nothing but reflecting light, bending time, following dimensions and rules which are beyond me. It's something that I was only just catching at before, something cold and exotic. It's rainwater, stream and groundwater, moving unseen through the aquifers with all their currents and counter-currents. The weather systems have cut through and sprung out of ancient bedrock, and formed shifting riverbeds; all the rivers out there, they're in me, under my skin, zinging through my muscles, moving the hairs in my follicles. The blood that flows in my veins moves with the same strange mineral energy as this body of water, and weighted by the earth and the moon, it pulls me with it.

From Bodmin I want to head up towards the highest point in the peninsula. I want to stand where the water begins and view what I can of England tapering away to a small fist punching at the Atlantic. I might go in over my boots; it's the saturated and boggy upper limit of the watershed, the spawning ground of the whole water system of this western edge. I want to know if otters would travel all the way to the source.

It's a few miles of winding lanes on the bike, and my route is plot-

ted; all I have to do now is pedal. I find it takes longer than I antici-
pated, due to the elaborate Cornish contours which somehow I didn't
notice on the map, but eventually I get as far as Camelford. Here, I
find a watering hole next to where the young river Camel bubbles
under the road. I need to recover myself, reconnect with my legs and
drink tea. I burst into a tea shop, red-faced and wobbly, and eat an
unladylike number of flapjacks. Before I leave, my aching muscles
need attention. Cafe owner and customers staring at me through the
window, I begin to perform some outdoor stretching before heading
onward.

Fully warmed up, I inspect the stream. It is greened by wide banks
covered in new shoots of wild garlic and tongues of flag iris. It riffles
shallow and clear over fans of pebbles and shale before disappearing
under a sturdy granite bridge. There is otter spraint here, and a shop-
per with a pushchair looks at me quizzically as I clamber back up the
banks. It would be tempting to follow this river, which is all easy
downhill, but today I want to go a little further uphill, to find the
beginning of the rivers that run south. The Camel River flows to the
other side of the peninsula, tumbling north through a lattice of woods
and farmland toward the coast. Where the fresh water meets salt, it
develops silty flats, becomes tidal, widening through meadows and
reed beds all the way to Wadebridge. Attracted to the family cycle trail
which leads from Bodmin to Padstow, visitors can pedal along easy
miles of the river. There are otters painted like emblems on the information panels, but the real animals' presence is invisible.

Where the river reaches the coastline at the holiday resorts, the water sparkles turquoise as it flows over sand, then deepens and curls into the sea. At the fishing port of Padstow, the Camel is clean and populated with sea trout and salmon. It seems there is little conflict here; there are plentiful fish restaurants and otters fat as butter have occasionally been seen playing in the water by evening pub-goers, even sharing the bounty alongside fishing boats in broad daylight. Witnessing this is living, vibrant proof that things are getting better for otters.

My route, however, takes me upstream to where farmland rises up to the grizzled contours of the moor. The way is signposted toward Rough Tor (pronounced 'Rowtor'); the map shows a long straight road crossed by a grid of thin blue lines. I set off again and this time find the going quite easy; the tiny road undulates gently to the beginnings of the water pathways I want to find. A little further on I can see the ground roughening to steeper slopes, and in the distance an imposing double summit is unevenly veiled in mist.

I can see that Rough Tor is not the top of the moor; there is an even higher peak just beyond it. Bizarrely, the higher point is absent from the road signs. This peak has a name which makes women young and old giggle: Brown Willy. While I pause to rest my legs and scour the map, another traveller stops to talk. He puts down his backpack and takes a swig of water from his bottle. He tells me he has been walking across the moor for three days, camping in a bivvy bag on dry spots in the lee of the hills. He can name all the peaks and landmarks, and thinks the name Brown Willy makes a nonsense of the landscape. The old Cornish name was not Brown Willy, but

Bron Wennyly, he informs me. It means 'hill of the swallows'. This sounds good on the tongue. We look back at the peak for a while before he picks up his pack and strides off. He must be twenty years older than me, but his thin frame seems to carry the weight as if it's nothing.

Bron Wennyly. Hill of the Swallows. Hill where the swallows go. It might have been what early settlers noticed about the place; they saw that it attracted birds. The swallow was associated with the return of summer. It was appreciated because it killed gnats and other biting insects, and because it represented spring, fertility and hope. The Romans thought that swallows were the spirits of dead children. Some people used to think that the swallows roosted all winter, hiding underground in mud, underwater, or disappearing like Persephone into the rocks when the weather began to turn.

If they saw swallows gathering before their autumn migration around the hilltop, and the following year noticed the birds' spring return, people may have imagined them coming back out of the rocks of the hill. In the eighteenth century the meticulous natural historian Gilbert White wondered if swallows hibernated somewhere. Since they were often seen feeding over water it had been concluded that they might be hibernating under the surface of lakes. White was sceptical about this. When it was suspected that swallows left for some warmer place, and always came back to the same nests, White speculated, in his classic book *The Natural History of Selborne*, that these birds had some kind of compass inside their brain. Now we know that migratory birds such as swallows have a magnetic sense: one of the things that guides them, apart from the sun, the moon, the weather and the stars, is magnetite inside their brains.

I, in contrast, have to look at the map. It shows that the long straight track to Bron Wennyly is crossed by numerous streams. This is what I am looking for, and I repeatedly pull up to park the bike for a closer look. Flowing between the contours at cross-purposes to the road, trickling subversively beneath, they offer another way through the landscape. Each one is crossed by a solid, ancient-looking granite bridge. If they have passed, otters will always leave their signs at these crossing points, so each one is worth a look. Sometimes a greener patch of undergrowth shows where the otter regularly leaves its spraint; I often wondered about this until I learnt that it was the nutrients from the otters' droppings fertilising the soil and producing rich growth in the foliage of the grass.

The human path speaks of generations of passing feet: hooves of horses, fathoms of sheep, farm work, mines, toil and worship. It leads directly to the hill, with a straightness of purpose that feels unmistakably useful and important. The gnarled walls, locally called 'granite hedges', are set back from the road's edge like an old droveway. They are made of massive boulders, some of them put here during the Bronze Age. The rock is hidden and clothed in ivy and lichen, and studded with roots of windblown hawthorn and willow.

Ahead, huge amongst the wild flatness of the moor, the strange outcrops of Rough Tor and Bron Wennyly could be the rims of volcanoes. As I cycle along I can smell familiar moorland scents of peat, sheep and desiccating grassland. The rumpled flanks of the hill appear bleached and the grasses almost golden after the winter. My vision focuses on the scattered rocks that form ancient hut settlements, field systems, burial chambers and stone circles. There was once ancient forest here, but when people settled they began to clear it all. A wider and wider sky appeared, and by increments the landscape changed into what we see today. Underneath and through it all, the otters are there, leaving only scent and prints that quickly fade.

I stop at each tiny stream that crosses beneath the road, scramble down the sides of the hunched bridges and look for signs of otter. The first bridge has an otter toilet right at its foot: a mossy rock covered in spraint. The spraint is dry in the sun but very obvious. The weather has not damaged it too much and it could be a day or two old. It is full of pieces of cartilage from some amphibian, perhaps a frog or a toad. So, they are here. Because I am moving through the landscape without the distraction of a car, the evidence is so much easier to find. As I hoped, otters must be coming this high up, right to the watersheds of the moor.

In my notebook I mark down my questions: does this mean that they are seeking refuge in these quieter places, or is it that there are more otters, and they are expanding their territory? The second bridge, a mile or two further on, has a flat rock beneath and there is a littering of tiny scales, bones and scent marks all over it. A mile later the road peters out and I park my bike; now there is nothing but a narrow track leading to the hill. A granite clapper bridge crosses the next stream. Beneath, pristine water meanders around boulders and pebbles. I clamber down to the sonorous underbelly of the bridge and wade about in the dark searching for more signs; there are two reservoirs nearby, one further upstream that is stocked with fish, and one downstream. Sure enough there are signs of otter again; there must be plenty to eat, even at this last edge of the winter, when food can still be scarce.

At 420 metres high, my guidebook says, from the top of the hill on a clear day you can see both coasts. Halfway up the first peak, a granite boulder the shape of a giant yoke is speckled and furred with lichens in a mottled skin of brown, lime-green, olive, slate-grey and black. Below, the early spring sun has formed a glassy heat-haze over the dry grasses. They wave like grass in a prairie, covering the hills in all directions. The air smells of a different moor. It has the strangeness of the sea and none of the sweetness of the heather and bracken on my patch on Dartmoor.

Mist curls over the view like smoke when I get to the top, and I wonder how the weather can have changed so rapidly. From beside a tall megalith, a piebald pony stands like a wild mustang, looking at me through a wind-blown fringe. This treeless moor is the source for many of Cornwall's rivers. It originally bore the name of the river Fowey, or *Fowydh* in Cornish, meaning 'Beech tree river'; the Camel, De Lank, Ottery, Inny and Lynher also spring from here. Again I wonder at the naming of places. Why is the moor no longer called after its main river, as Dartmoor is, but instead connected to the human settlement of Bodmin? This name is from the Cornish *Bos Venegh*, meaning place of monks. It seems as if over time the name has been masked and the elemental essence of the place forgotten.

A raven swings into view, his wings carrying a weird blueness about them. The raven calls. His gravelly voice contains a coarse grain, like the grating of granite on granite. He turns and soars, the wings stretch black tips to catch updrafts of air. As he flies over my head I catch sight of a curved black beak and a ruff of plumage that ripples in the wind. He tips toward me, lands near and folds his wings in perfect symmetry. He watches me like a shadow, a spark in his eyes, something about him drawing the darkness from my unconscious. I eat my snack uneasily while he waits, and when there is food left over he watches for me to move on before alighting to inspect it.

My eye dawdles over the marshy tilt of the valley bottom. This is where the rivers begin, and it's boggy and famous for swallowing people. Perhaps the raven is waiting for the ground to leave him my bones. When Daphne du Maurier first came here she fell in love with the mixture of swamp and ruggedness. She wisely explored on horseback, and for her the desolate scenery and waters coursing down to the sea were like a portal to another world. In reality she got lost and struggled in the marshes here, took shelter in a ruined cottage during a rainstorm and later had to turn back to Jamaica Inn. The adventure was one of the inspirations for her famous novel, and writing about how gruesome the moor could be on a wet day, du Maurier said that she came 'unprepared for its dark, diabolic beauty'. She relished the 'frowning tors and craggy rocks' and delighted in the wildness of the landscape and its people. In *Jamaica Inn* she embellished them further: 'mile upon mile of bleak moorland, dark and untraversed, rolling like a desert land to some unseen horizon … the very children would be born twisted, like the blackened shrubs of broom, bent by the force of a wind that never ceased … Their minds would be twisted, too, their thoughts evil, dwelling as they must amidst marshland and granite, harsh heather and crumbling stone.'

Having read du Maurier's book before setting out, it might seem better to turn back well before dusk. But I remind myself that the four winds she describes blowing in all directions in fact mean that the air is pure and pollution-free. The endless misting Atlantic rain feels refreshing after my cycling exertions, and the moor has no real malevolence. It is a place that has visibly supported generations of ancestors; from the slopes I can see where two thousand years ago people laboured in the open air to uncover alluvial ore hidden in the layers of gravel in the beds of the stream. You can almost hear the axe-split and gravel-pour echoing in the sounds of the birds, the wind, the water and the wet granite. Horn picks and wooden shovels would have been used to dig down to rich veins in the bedrock. The ground is riddled with dips and gulleys; the sides of the wide valley dotted with hut circles, standing stones and mine remains. The ground has been scarred and healed, and all around streams run downhill like veins of energy.

I begin the ascent of the second peak. In the distance to my right I can see the tower of a granite church, some stunted trees and a farm; there is something ancient about their form, as if they occurred as naturally as the weathered rocks of the summit, where the formations are sculpted with intimate shapes: giant eggs ready to hatch, stacked towers striated with claw marks, strange limbs with missing mouthfuls, hidden caves and narrow, water-filled arteries. All these natural megaliths seem to be the bones of Cornwall, the solid core of something immutable. Muscular contours of peat and grass texture and colour the valley like deer hide, shifting with the movement of cloudscapes. Far below, the rainwater that gathered up here has run all the way down to meet the sea. It soaks into the sandy-edged epidermis of the coast, through shifting estuaries and into the tides.

There is something unusual about the wind around Bron Wennyly. Du Maurier's words capture the way it moves through the stones: 'Strange winds blew from nowhere; they crept along the surface of the grass, and the grass shivered; they breathed upon little pools of rain in the hollowed stones, and the pools rippled. Sometimes the wind shouted and cried, and the cry echoed in the crevices, and moaned, and was lost again.'

Coming down from the summit I enjoy the shiver of it all, not really believing but not lingering either. I follow a stony path to the beginnings of the trickling De Lank River. It is all grassy whispering between banks and layers of black peat. I kneel down and test the temperature. It is cold, very clean and tastes only of spring. There is an abiding smell of fox, but no otter. Animals can obscure one another's scents, and with a fox here it will be hard to detect anything else.

Here and there the flow syrups into a pool in trails of bubbles. I take off my boots and socks and measure the water with my toes; after the first shock, every molecule of my blood is pulsing with the cold current.

Deeper onto the moor my eyes adjust to the reflected light from pools, the meshing of sphagnum mosses, drooping lichens and marsh grasses. I startle a hen partridge, which scuttles away and vanishes into a maze of browns and greys. Amongst it all, I find the hidden leaves of primroses with tiny flower buds.

There are more ponies grazing. They are every colour of the moor: granite, buff, speckled, chocolate, black. Ankle deep, one of them drinks and then looks up at me, startled, its lips dribbling ribbons of water. A snipe flies up with a sharp cry, its long beak and mottled wings a flutter of light and mist.

The bog has made a nonsense of time. It could be hours that I have been walking through the sodden peat and wind-sound. The path is lost. In the sun, the hills are the colour of a lion's skin. A wheatear lifts out of the grass and bounces away, and somewhere above larks are busy chiselling tunes into the sky. The water moves everywhere in my senses, in the splash of my footsteps, in the misty sky, in the black pools. Otters have been here, following eels thick as bike chains, catching at bright, wet frogs that have come to lay thick masses of spawn, or find them hiding amongst the rushes and tussock roots. I follow what could be a mammal path, where grass has been smoothed in a trail around a pool. There is no spraint, but one paw-print in some silt, as if an otter began to appear and then changed its mind.

In the end I find the source because of the way it catches the falling light. The peat is a slick scattered everywhere; wet pours over my boots, and through the clear water, a strand of glass spins out of the ground.

A curlew forms itself out of a granite boulder. Its cry makes the hair on the nape of my neck come alive. I can't stay to listen for otters at nightfall, even though I can tell they've been up here. There is plenty alive to attract them to these pools, and maybe if I lingered I would hear the whistle of an otter, but it's a long way back to the solid world, and the light is failing.

From just north-west of Bron Wennyly, beneath Buttern Hill, the river Fowey gathers itself and spills off the moor. It shoulders and slumps round boulders, carving steeply into its own valley, riding round bends, soaking into reedy marshes where elvers and frogs thrive. These marshes are as important to otters as the river. They purify the river and provide rich feeding grounds. The water crosses under the roaring A30 dual carriageway, moves through a drowned valley, then into miles of roughly walled moorland fields. Further on, in steep-sided places, it cuts through sheer rock, escaping quickly from the moorland, and flows hidden into a maze of mossy boulders, patches of withy and knotted hazel woods.

On my bike I freewheel down a quiet backway which follows the Fowey's long path off the moor. Where the road keeps company with the young river I can investigate easily. I don't have to get my feet wet or expend any more energy on my aching leg muscles. I can whizz down alongside the water, feeling the cold right through me and the rattle of the road in my skeleton.

This gravelly part of the river is perfect for otters. It's an important spawning ground for sea trout and salmon and is protected from fishing. The fish need shallow gravel or shale in streams where they can lay their eggs. I wonder if otters would eat fish eggs? Perhaps they would eat those creatures that eat the eggs. When I stop I smell spraint from the road. I pause beside each bridge and inspect for signs. At every obvious point there is recent evidence that otters have passed through. Some of the spraint is only a day old and still firm, without being dried into a splinter, and some of it is very dry, perhaps as old as a few days, and deposited regularly all along the river at intervals of between ten and fifty metres. This could be because the river is shallow and the otter comes out frequently to eat and scent-mark. There could be numerous other reasons; there could be a resident otter that wants to keep this patch to itself. I have found this type of frequent marking on other small rivers; on larger waterways it can be harder to find signs, and sometimes I don't find any at all until I look at the tributary. All this, as usual, leaves more questions than answers. All I can say is that the otter here is eating plenty of fish. The spraint are small, as otters eat tiny meals, foraging and grazing as they go; there is never time to stay still for more than a moment. This one may be an adult dog otter, as I've heard the larger the otter, the smaller the spraint; cubs can leave the largest, smelliest signs of all. As with the signs I found high on the moor, it's clear the otter is an opportunistic feeder and will eat almost anything that appears edible.

There was a fishery near to Rough Tor, and faced with a still-water pond full of fish, the otter will instinctively kill, and return another day to make a new catch, so proper measures are vital unless you want all-out guerrilla war where the enemy is not visible and nocturnal stealth is its special talent.

A little deeper now, the river swirls and eddies into all the places eel, pike and brown trout can hide. It glitters past people's gardens, past their car ports, lawns and trampolines. I stop, lean my bike against an oak and sit down on the bank; the surfaces are all mossy boulders and warm granite. The river is hardly managed here; the few bridges that cross it seem barely used, and nothing is put on the fields. The river is healthy and self-regulating; it is an inviting home for wildlife and maintains itself without interference. Up and down the banks are the complex root systems of ash trees, which otters particularly love to use as holts as they provide hidden shelter and easy access to the water. I find otter spraint almost every hundred yards. There might be more than one living on this stretch as food seems to be plentiful and housing abundant. I have no way of knowing for sure how many there are on this river; it may be two bitch otters whose territory overlaps, with an occasional male visitor, but probably not more than this. The inhabitants of the few granite farmhouses that exist along this stretch might be aware of otters in the river, and their lives will be richer for it. The presence of wild animals feeds a hunger, appeals to the imagination, carries our fascination.

A few days later, when my legs have recovered enough to walk, I take the car and visit the popular shopping venue of Trago Mills between Liskeard and Bodmin. This is not for the purpose of shopping but to nose a little more around the river. Here the river is leaving the granite of the moor, and in places the noise of the maturing water becomes tremendous as it crashes steeply southward. The water tumbles over rocks and through woodland; it meets with other water, races under roads and bridges, the moist spray feeding ferns and mosses on the banks. Otters come out of the water where it is fast moving and move stealthily through the rich undergrowth and rocky contours where nobody would notice them.

Just off the A390, people stop at Trago to shop on their journey west. A walk has been constructed along the clear waters of the Fowey; the flow of river is joined by another, and they slow where you can admire the flock of domestic chickens and ducks, or exercise your dog and take in the shops. Just outside the boundary of the shopping centre, not far from where people park their cars, there are covert otter-happenings. A confluence of two rivers coming directly off the moor means otter territories might overlap, providing an important place for otter interaction. I cross over to see what I can find out, and look at the bridge where the Neot River tumbles off the moor and meets the Fowey. It's a climb down a five-foot wall that has been built to support the bridge and protect the road from the river. Then there is a battle through brambles and nettles. Perhaps most people wouldn't bother with this, but if you have long trousers and a stick you can go looking for these otter signs, although you might need wellies for wading across the shallows in search of spraint.

Down at otter level, a whole world opens up. Beneath the bridge there is a fragrant mass of eroded silt which the river has carried downhill. It has piled up smoothly into a perfect otter service station, where the wet sand has preserved a hundred or more prints.

I spend some time trying to identify the number of otters that could have been here, but there are so many I cannot distinguish one animal from another. There is a rock with a heaped tower of fresh spraint and scatterings of many old remains. The site is used regularly. A recently half-eaten duck lies in the shallows. This was presumably taken from the ornamental Trago flock, as its remaining feathers are white. It must have been a great meal, and most of the bird, bar the wing bones and some feathers, has gone. Otters will eat birds; feathers are often found in spraint in the spring when waterbirds have young. In winter they feed on carrion when birds die of cold; starling feathers are commonly found in spraint near roosting grounds in wetland reed beds.

Canoeists have been riding the fast-flowing current further down-stream. In places otter entry points in the banks have been eroded by canoes. This may be one of the reasons otters are so evident higher upstream, but I can't assume they've fled. It is possible that they're not bothered by the canoes and carry on with their business at night when the noise and excitement have passed.

Eventually a shallower gradient levels out and the current is more graceful; the river is in its mature stage. Here the rock is Devonian slate; the water is just as clear but deeper and slower. Trees thin out and the land use changes to fields and farms, and villages and houses dot the rolling countryside. Downriver the first hint of an estuary begins. Creeks sidle into the main river, hidden tributaries lined with tall ash and beech trees carry themselves into the green fathoms of the Fowey. In places the water seems to absorb all the bright newness of the first leaves sprouting from their buds on the trees. Slowly the water changes. In places where the light has not yet reached, it seems nearly black, but in brighter stretches it turns to a deep bottleglass-green, the kind of colour you only find in rivers and lagoons salty with tidal water from the sea. All the hurly-burly of rushing off the moor has gone. This could be a different river entirely. It is half-sleepy, brewing secrets of sea trout, eel and bass which hide their clean, silver, ocean-going selves in the inscrutable depths.

I have been walking for an hour, but it's early and everything is still cold. Where there is no shelter there is a fragile tissue of frost; the river has edges of ice where it curves into slow pools. I catch a gleam in the current, and find early-morning ice crystals melting in a shimmy of tracks across the sand. An otter has rolled here, leaving an ample round squirm in the ground. Perhaps a pregnant female, she's hefted her weight back into some holt now, waiting for daylight to pass. There is no sign of which direction she took. There is a delicate, feminine light coming through the branches; a hidden-ness in the buds of willows, in bright green catkin tails and the blood-red of dogwood twigs. Where the sun touches the water, twists of steam rise and billow in frail breaths.

The bridges become grander as the Fowey grows in maturity; the six magnificent medieval granite arches at Lostwithiel host gravel banks of debris that have been washed down from the moorland miles upstream. Nestling in the folds of the river valley, this small town was once a port that exported tin, wool and leather. The bridge is the last crossing point – or the first if you are coming from the sea – although the town is far enough from the seaside and its attractions to remain quiet and unspoilt.

The river is a wide glassy expanse, reflecting flattering patterns of sky and houses. There are more signs of otter beneath the bridge. Otters living near towns are less startled by human noise and detritus and will retreat to a holt near the water during the day, carrying on at dusk in spite of the recent daytime scents and mayhem of cars, lights, bonfires, fetes, canoeing races, dogs or paddling families. It could be that otters live closer to us than we think. Keeping a low profile comes naturally to them, and even if they were out in the day, we may not catch sight of their subtle contours slinking into bankside vegetation.

Beyond the town the river has reached its inter-tidal zone. Silt and muddy banks should make it easier to find otter tracks, but reed beds prevent me exploring further. The tides dominate the river's landscape here. Sea trout, bass, flounder and mullet can be caught at certain stages of the tide, and a little further seaward, salt marsh and mudflats spread uncertain depths of stickiness. This is off-putting for people, but not for otters. As the river estuary broadens, the Fowey mixes with the confluences of the river Lerryn and countless other sinuously green and wooded tributaries and creeks; pleasure boats sail up from their moorings in the deep harbour of the seaside town of Fowey. The landscape becomes more and more shaped by human settlement and engineering, and the sounds, smells and distractions of the seaside take precedence: fish and chip shops, riverside pubs, houses and restaurants, all deliciously enticing to people, but not to otters. Where the river widens to its mouth, all eyes are on the water, from windows, balconies, verandas, and look-out points. The yacht club, coast guard and large cargo ships make use of the forty-foot depths; otters will slide past in the half-light, disappearing to the quieter creeks and bends in the river where they cannot be seen, and any evidence is rinsed out on the tide.

While humans look out from scenic windows stacked up the hillside, the otter becomes invisible again. There is something mythic about this disappearing; it sinks back into the marvellous, fantastical underworld of its own element, where we can't follow, and the banks and edges of this river seem full of unreachable wonders: a rain-coloured heron; a wriggling pool of frogspawn; green tongues of flag iris pushing up through the wobbly textures of soaked moss.

It's possible to come upon this scene and see it as if it were the only river you'd ever set eyes on. In Kenneth Grahame's *Wind in the Willows*, Mole, worn down by the dirt and drudgery of spring-cleaning, decides to go for a walk and encounters the river's sudden strangeness: 'Never in his life had he seen a river before – this sleek, sinuous, full-bodied animal, chasing and chuckling, gripping things with a gurgle and leaving them with a laugh … All was a-shake and a-shiver – glints and gleams and sparkles, rustle and swirl, chatter and bubble. The Mole was bewitched, entranced, fascinated.' I can imagine Grahame himself walking here, equally enchanted by the water: 'he sat on the bank, while the river still chattered on to him, a babbling procession of the best stories in the world.'

Far to the west of Fowey is a place where those stories seem very close to the surface. Almost at the tip of the peninsula, where the land of West Penwith knuckles the Celtic sea, the human, the industrial and the natural rub up next to one another, and otters squeeze between. The dominant feature is the granite, which has been used by humans for thousands of years, at first to build carns, walls and chambered tombs locally called quoits, and later to fashion the solid grey cottages of the old mining and fishing villages. The uniqueness of this small area is that it looks like a single granite and moorland outcrop, bordered on three sides by the Atlantic. In fact the bedrock is joined deep underground to the rest of the granite in the West Country. West Penwith has a special atmosphere because it is the only granite intrusion in the west to be in contact with and eroded by the sea. Its high heathery moorland and patchwork of field systems are dotted with vestiges of ancient people, giving a wild and mystical aura to the landscape. Standing stones litter the fields and moors, many of which have remained mysteriously inaccessible and have never been excavated or explored. The numinous peninsular light attracts people here in droves, to visit the beaches and porths, walk the cliffs and explore the Bronze Age remains.

Between the eroding granite cliffs and huddled fishing towns is an isthmus which connects Penwith to the rest of Cornwall, and along this is a tidal estuary which feeds a pocket of pristine dunes and rare grasses. This is the Hayle Estuary. All around it is a web of roads, houses and seasonal tourist resorts. In this tidal zone people and sea ebb and flow. Different worlds meet and overlap. Twice a day, beside the road, steady ground becomes slick sand, firm becomes fluid, and the sea pours in. What delineates the boundary between land and sea becomes meaningless, and otters come to forage in the rich feeding grounds of these edges.

As the pattern of the otter population re-establishes itself and reaches its original levels, people report seeing otters swimming just off the coast. Potentially the whole of the water system is at the disposal of this animal. Its versatility must be why it survived in spite of all that humans have done to encroach on it.

From the cafe in a converted branch-line station close to Hayle, if you know what to look for, you can watch a head rise like a periscope, as a mother otter guides her two cubs quickly across the brackish water of the estuary. She may haul out on the strip of beach alongside the road on one side or the railway line on the other. Here she can teach her cubs to fish, and watch them tussle on their own with a crab or small eel.

It may look like a pack of otters, but usually it is a mother and sub-adults. Otters are unique among carnivores for the length of time they care for their offspring. When mating is over the male usually leaves the female to care for any subsequent cubs, and she will remain feeding and teaching them for up to twelve or even eighteen months. By contrast, fox cubs mature and leave from six months of age, and badgers mature at a similar rate.

Otters are impossibly secretive about their natal holts. After the cubs are born, the mother otter will slip invisibly in and out to feed herself at night. This is a covert operation for which she is perfectly designed. With her camouflaged, streamlined form she is almost impossible to see in daylight and totally invisible in the dark. When she has small cubs, she is constantly on the alert. She spraints in the water to avoid detection, perhaps by adult males. Nursing bitch otters may hide their cubs to protect them from being killed by the dog otter. Recent reports of cannibalism amongst otters in south-west England confirm that litters of cubs have been killed and eaten by a male who was subsequently killed on a road. Although this seems tragic, it at least suggests that otters are becoming more numerous. If populations are reaching healthy levels, or even capacity in some areas, numbers regulate themselves in this grisly way.

But why do otters, unlike many other carnivores, live with their young for so long? The answer lies in the otter's environment and feeding habits. The skills required for hunting underwater are huge and the process of learning is long-winded. A mother otter will help to feed her young far beyond the age at which they are weaned. After two months, while the cubs are still suckling, the mother may begin bringing the cubs small fish. At this point she may carry or lead them out of the holt and down to the water. Here she can teach them to swim, often by bodily immersing them, in spite of pitiful protestations. The furry cubs are highly buoyant and their thick, fluffy fur traps air and helps them to stay afloat, but can make swimming difficult. Most learn to float quickly, and in a few days will be confident enough to follow their mother into the water and float about close to her, gradually learning the skills for swimming.

During the first few months of their lives, as they grow, the cubs follow their mother everywhere and wait by the water, whistling for her while she forages and brings them food. The family are rarely out of one another's sight. At this point their hunting skills are still non-existent; it is only when the mother deliberately brings her young live prey that they can begin to practise their skills. Young otters will watch their mother closely while she is fishing, and eventually after a few months may try to catch something for themselves. With slippery prey, this can be an awkward business. If they do make a catch it will be something slow-moving – for example, crabs or crayfish, which have low nutritional value and a painful bite – but it's a start. It may take many months before a sub-adult will be able to hunt more skil-fully for itself, but gradually it will learn through the same process of trial and error as all mammals do. Eventually, by nine months to a year old, it will have the beginnings of the skills it needs to survive in the challenging environment of marsh, river, estuary or coast.

After this the female will take off, and sometimes the young adults are bereft, calling and searching for her for weeks. The American sea otter cares for her pups to an even greater degree. The marine envir-onment is more challenging than British shores, and the sea otters have developed astonishing bonds with their young. The mother sea otter floats out at sea all the time, carrying a single cub clutched to her chest almost continuously. As the cub grows, mother and young still hold on to one another constantly in an intense display of affection.

They clasp their front paws around one another's necks, or link paws as if they are holding hands. If they are ever separated accidentally they scream and wail until they are reunited.

The riverine otter in Britain does not have the same abilities as its translantic marine cousin; it lacks the clutching primate-like forearms of the sea otter. If she absolutely has to, a British otter will carry her cubs, but she will take them one by one dangling from her jaws. Out in the world of swift currents and busy roads, there is not much a mother otter can do to rescue her cub if it gets into difficulties. There is no holding hands, nor any safety harness for toddler otters lacking road awareness.

Just before my visit, one such incident was reported in the news. In the sodium glow of overhead lighting on the causeway at Hayle, a young otter mother slipped out of the water and galloped over the road. It was rush hour, and just getting dark. The driver of the car that hit her may not even have been aware of what had happened. The otter was found soon afterwards by Sam Lawley, a local vet driving home after work. At first Sam thought she had found a stray dog, but soon saw that it was a wild otter. She could see that this otter had been hit and killed instantly, and on closer inspection, turning the body over, she found that it was a lactating female.

The causeway was built so that people could cross the river safely; in the past many lives were lost as people were guided over the treacherous sands of the estuary. The road made life safer for people, but not for otters. Human engineering has thoughtlessly cut into otter territory a million times in this way. Now the road has expanded and grown into a dual carriageway; if you drive along it you could be forgiven for not noticing that it crosses the river Hayle at the moment it pours into the estuary.

Close to the mother otter's body, Sam Lawley could hear a strange, high, whickering cry. Following the sound, she found it was a very frightened baby otter, hidden in the grass. Crying for its lifeless mother, it refused to leave her body. With some difficulty, the vet carefully picked up the cub, wrapped it in a jumper and took the reluctant, wriggling package to the local wildlife park. There it was checked over and nursed back to health. Another cub was found later close to the same spot, but it had not survived.

The luckier cub was named Harry, after the escape artist Houdini, because it quickly did what otters do. One night under cover of darkness it squeezed through into a neighbouring pen to investigate the smells of the other otters and raid any remains of their food. The bewildered keepers did not find Harry until hours later, a hidden brown shadow huddled against the brown walls of the pen, almost completely invisible.

At around six months old, otter cubs are not sufficiently independent to survive. Harry would need to be kept for at least a year and then released, according to guidelines, close to where he was found. This method sounds sensible, humane and forward-thinking, but the releasing of such otters is not always supported. Wildlife experts disagree on many issues surrounding otters, and releasing orphaned

186

otters is not necessarily seen to be as helpful as it sounds. Otters are territorial, and if an adult otter disappears, as this mother did, the population in the South West is sufficient that another adult could quickly take up residence in the vacated range. Before the young newcomer can settle into its new home it may be attacked, driven out or even killed by another otter in the area. Post-mortems reveal that one cause of death can be from infected bites; this may be from territorial fighting with other otters. If new otters are introduced where a balance has established itself, we could inadvertently be causing trouble.

The alternatives, however, are not generally advisable. If an otter is introduced to another area it may carry disease there, or may arrive in a river where there is not a sufficient fish population to support it. Another alternative is to keep them in captivity, which is costly, difficult and unkind. Finally the solution may be to have these cubs put down, but this option is not usually considered if the cub is healthy.

So many otter cubs were found abandoned or orphaned throughout one flooding and freezing winter that the Secret World wildlife sanctuary in Somerset was almost overwhelmed. One after another, cubs were brought in, all between two and six months of age and all with big appetites. The fish bills soared; since they have to keep the otter cubs for at least a year, until each one is mature enough to survive, the cost ran to thousands of pounds.

Because of the recent otter death, I plan to walk up the river Hayle as far as I can and try to find out more about what happened. I want to know why this otter might have needed to cross a busy road. I'll survey for signs of any remaining otters or at least find out if new otters have made their presence known. The dead otter was found at the road on the edge of the estuary, less than half a mile from where the fresh waters pour into the Atlantic. These otters might have been feeding on the rich source of marine prey provided by the tides and the sea nearby. If so, there may be a healthy and well-fed population.

I make the journey by train, planning to stay for a few days in a friend's house in nearby St Ives. As the main-line train swings inland through the Cornish broccoli and turnip fields, there is little hint of the seascape about to unfold. Finally, the one-carriage train from Penzance to St Ives rattles into Lelant, opposite the old river port of Hayle.

Heyl, in Cornish, means estuary. Tin and copper were discovered here and mined thousands of years ago, and later boats came to trade. A few hill forts and homesteads scattered the moorland which gave way to the wide mouth of England's most westerly estuary. The estuary was wider and deeper then, and fishing boats could come upriver as far as what is now the town of St Erth. The industry grew, a port developed to land coal, and trade flourished. Fragments of pottery, pieces of amphorae, vases that contained olive oil or wine, have been found. Christian monks built churches and the harbour expanded to deal with more trade from the growing mining industry. Today that industry has declined. Remains of old mines, the foundry, the harbour, and shops and workers' cottages from the industrial age still line the sea front, but the tourist industry now dominates.

The river Hayle is only twelve miles long from source to sea. It begins in rough meadowland and flows west, passing through a wooded granite gorge before turning north and meandering through fields and villages toward a tidal barrier; here it disgorges into the wide Hayle Estuary and salt marshes of Lelant. Just before it meets the ocean, it flows between the mountainous white dunes of Porth Kidney Sands and Hayle Towans. Suddenly the muddy brown estuary becomes a splendid azure-blue, mingling with a brilliance that seems to come out of fairy tale. Finally it pours over an enormous white sand beach, laced with miles of salty foam and seaweed. Otters run to and fro amongst it all, presumably under cover of darkness. They are very rarely seen by humans, and not generally known about, except perhaps by the other margin dwellers, the curlews, egrets and shelducks.

My first job is to find the spot where the otter was hit and do some detective work; why was she crossing the road and not going under it? Is there anything that can be done to prevent this happening again? I notice that the road does not seem very busy. Most of the traffic has been diverted onto the A30 to Penzance, which is raised high on a viaduct. On the other side of a low wall a clear freshwater stream runs into the Hayle Estuary; it has plenty of muddy banks, no litter, swirls of green weed and little minnows darting around like arrows. Running alongside the stream, just below the wall on the soft marsh grass, a well-defined otter path runs safely parallel to the road. I find large prints entering the silt and plenty of tidal fishing ground beyond. But when I get to where the river Hayle itself enters the estuary it is a different story. The river is channelled under the road in a series of disastrous concrete tunnels and the water is rushing through narrow grille openings at a speed off-putting to any otter. I cross the road and find the next problem: the Hayle tidal barrier. A metal barrage blocks the flow, and the large volume of water is controlled by a mesh of fencing and tunnels. It is a maze of noisy, rushing water, slippery concrete and bars of metal.

There are some concessions, adjustments humans have decided would make it easier for an otter. They are entirely logical to us, but we don't know how an otter would see it. Steps have been built up to a gate with a small built-in wildlife exit which is exactly otter-sized. There is no sign that any otter has been tempted to use this facility, perhaps because just beyond the gate, otter mesh prevents a road crossing. The otters still want to get to the estuary, and since otters are determined and economical they will take the straightest way there, ignoring any sinuous safety passages that have been provided. This seems to be what happened when the mother otter at Hayle brought her two cubs to the estuary. Perhaps it was their first trip to the big fishing ground, or perhaps this night they were just unlucky. The mother avoided the tidal barrier completely. Instead she must have exited the river upstream, led her cubs under the barbed-wire fence, trotted with them both in tow to the five-bar gate, slipped easily under it and ambled together in a family group over the road. But on the other side of the road, the otter meets the wall, and has to travel quite a distance on the tarmac of the road with two dawdling cubs. Perhaps they were distracted and meandered a little away from her, or one of them was left behind. In any case their journey to find a way down to the water of the estuary took too long and they never reached their destination.

Sluices, engineered riverbanks and tidal barriers are not otter-friendly. But other wildlife is affected too; I think about the salmon. Travelling upstream to spawn, a salmon can no more change her plans than an otter.

The reed beds beside the river Hayle on the narrow sliver of land upstream have been acquired by the RSPB; wedged between the fast roads are two ponds which provide an attractive island refuge for migrant waterbirds that visit the estuary. The gate into this land has otter-proof fencing to deter otters from crossing to the road; but the other side of the river must be privately owned, so there is no special fencing, and the otters can cross freely. There is no road sign to slow down. Whoever had been driving in the dark that night may not have even noticed a soft thump under their wheels.

Upstream, the Hayle River has been extensively managed. The water flows between built-up flood banks and a mass of reeds. In spite of the artificial banks, the reeds dominate, and provide excellent cover for otters. You can walk as far as the village of St Erth, where in Roman times the seawater used to reach. I arrive in early March sunlight. Warmth radiates from the local granite of the houses, the sturdy bridge and the old church tower. Daffodils are already starting to come out, their yellow trumpets blazing in a cold east wind.

There's no sign of otter around the old granite bridge, even though it has been fine and dry for more than a week. Over the bridge the river looks like it has been straightened, the banks built up so that there are two levels, a lower, rather water-logged one and a second, higher ridge where you can walk and survey the water and the surrounding fields. A wide grassy area provides a walkway alongside some coarse-fishery ponds. I eye these warily through my binoculars, having recently learnt that some coarse-fishery owners have resorted to snaring otters who take their fish. But both ponds seem as if they are fenced and fairly well protected from predators. All around the edges of the ponds are benches, and one or two fishermen are already installed, at perfect peace in the sunshine, meditating on the still water and the swaying forest of reeds.

Small fishes flash about in water that is startlingly clear, and weed is flowing in graceful fronds. We are within sniffing distance of roads, tips, industries and recycling centres, but this area is rich with clean springs. They dot the old Pathfinder map I have, and flow into channels, disappearing under ground and appearing through drainage pipes that pour them into the river.

Before the drains, this ground would have been marsh. Willow and rushes growing thickly all along the banks attest to this. I scan for signs of slides or sprainting spots but there are none. It doesn't seem right – this river is perfect, glittering and clean. It riffles and swirls and eddies under steep banks. Sometimes it is shallow, sometimes quite deep, but never more than five or six feet across. Although the willow and hazel provide some cover, the banks have been altered, and reinforced with rocks and wire. Everywhere are dog prints, and in numerous places the banks have fallen in where walkers have allowed their animals to play in the water. It is a half-wild river. I begin to see how this landscape is balanced between wildness and domestication. The earth has been moved, the currents shifted into pipes; the water pours with an unquenchable vitality, but there is an absence. This stream is bereft of its lucky spirit – the bitch otter and one cub are dead. Otters are like a live wire. Without them, a river is an empty version of itself. One otter may be returned a year from now, but its chances of survival are small. It will understand no more about the road than its mother.

Finally I find a footbridge. Bridges are good news for an otter tracker because they encourage the otter to scent-mark. Beneath a bridge the spraint is protected from the elements and it can be found by anyone or anything sniffing about. I slither down, aware that the concrete incline is slightly steeper than I can manage. When I am fully beneath I can see something unmistakable tucked in an impossible gap up between the ground and the underside of the bridge. I claw my way back up and push my face in close: dried spraint! I pick it up and take a sniff. Perfume. How do otters manage to get it to smell like this? It's sweet, like jasmine tea with a hint of bergamot. But it crumbles to dust in my fingers – I know it could be three weeks old; perhaps one of the last marks of the dead otter. I am only two miles or so from where she was killed, so within her range. There are tiny bones and scales inside it. A last supper? Then something else catches my eye. Strangely, there is another dropping, but not otter. It might be mink, or even polecat, but it is newer, longer than the otter spraint and more sculpted, tapering to a point. It has a different sort of smell. I have heard of wild mammals 'over-pooing' in popular spots, a little like the dog-and-lamp-post scenario, as they compete for territory, trying to get rid of the other's scent. There is a biological sense to this.

An otter will spraint to lay claim to this particular feeding ground, so that it will have enough food for itself, other otters will flee and stocks will not become depleted.

I turn and climb carefully out from under the bridge. On the other side, there is more spraint, this time closer to the water, an otter on its way back to its holt perhaps. Further on there is another bridge to look at, and I find more spraint, but this is fresher. Otters must choose bridges because they are prominent and offer a sheltered spot where the scent mark can last longer. Further on I notice a strange greyish mass on the opposite bank and scramble across to see. A poke with my stick turns up two freshly dead and half-eaten frogs. They are still moist. Otters will skin frogs and toads to avoid ingesting any toxins, and this must be what has happened here. This is confirmation of another, possibly new otter in the area.

As my foray progresses upriver something strange happens. My heart lifts more and more at every sign or spraint that I find. Soon, each clue seems to sparkle, as if I am following a treasure trail. Each crinkly sculpture is thrilling proof of live otters moving into the territory left by the others. The signs are like lost artefacts, carefully crafted clues; every little half-digested bone, beetle or scale brings me closer to the confirmation, the reality of a new otter.

After I have been going for more hours than I can think of, I find a village where the granite bridge is too low to navigate. This is Relubbus, about six miles from the estuary. The river is a narrow stream now, and on the other side of the bridge it begins to rise through a steep wooded valley. I walk on through mixed beech and ash trees, but the water is rockier and shallow and it is harder to find any more signs. This is not far from the source; it must be only a mile or so away. Nearby is the granite high ground of Trescowe Common, where a gouged labyrinth of disused tin mines lies beneath the surface. I find a last low bridge and look through a dark mossy tunnel beside where the water travels; perhaps it was built for storm water, but it smells distinctly ottery. As my eyes adjust to the darkness I can see that a mother and at least one cub have left their signs. One spraint is surprisingly bulky, dark with fish-matter and not very old. On the other bank somebody's garden lawn glistens neat and green. A little tunnel goes through the wild undergrowth and up the bank. There is nothing to stop an inquisitive otter from trotting into the garden. In fact every indication is that this is exactly what the otter has done. I wonder if the householder knows who has been there at night? Perhaps they will notice only that the fish in their pond have vanished, or perhaps they have given up restocking long ago, confused about the covert fish thief, blaming it on the local herons.

Another look at the map reveals that the source of the river will be hard to locate in the dwindling sunlight. I realise I'm hungry, my bones ache and I still have to get back to the train station. The post office in the village is closed and there won't be a bus until hours later. I turn and follow the current back downstream.

I return with soaring spirits. The new otter (or even otters) who left these signs could be miles away in a holt on another river, but could just as easily be sleeping on a couch somewhere close. They could be about to wake and begin their night. The tussock grasses in the thicket on the opposite side of the river look inviting; no walkers or dogs could penetrate here. The mounds of grass are protected by a dense wall of brambles. In places the grass has become a mass of deep cushions, bunched so far with plumpness that if an otter were in there curled up asleep it would be impossible to see from the outside.

When I stand and look down into this small river I can see a whole odyssey of meanderings. It's in the hair-like fronds of weed, the reflected over-hanging wildness of gorse and heather, the lichen-drenched willows with their tips touching the water surface and the swirling eddies in the deeper water. I have half a mind to fall in. Just to feel the water flooding over my head, like one of Charles Kingsley's water-babies. The attraction of diving into this sparkling realm must be why it conjures so many magical creatures, and perhaps why otters turn up so frequently in stories. In *The Water-Babies* the appearance of an otter has all the enchantment and dual nature of some mysterious riverbed being. But this otter is not at all benevolent. In the story, the water-baby Tom sees a family of otters 'who were swimming about, and rolling, and diving, and twisting, and wrestling, and cuddling, and kissing and biting, and scratching, in the most charming fashion that ever was seen'. Kingsley must have observed otters to be able to describe them like this, even in such an anthropomorphic way. 'And if you don't believe me,' the author continues, butting into the narrative as if to drop in a comical note of authority, 'you may go to the Zoological gardens (for I'm afraid that you won't see it nearer, unless perhaps, you get up at five in the morning and go down to Cordery's Moor, and watch by the great withy pollard which hangs over the backwater, where otters breed sometimes), and then say, if otters at play in the water are not the merriest, lithest, graceful-est creatures you ever saw.'

When the mother otter spots Tom the story takes a sinister turn. She immediately decides he is prey, and calls her children to hunt and eat him. I remember the horror of reading this as a child. The whole family of otters circles around Tom with 'wicked eyes'. But as the predators surround Tom they notice that he has hands, declare him an 'eft' and leave him alone because he obviously isn't edible. I've still never found out what an 'eft' is, but it was looking like one that got Tom off the hook and children around the world could breathe a sigh of relief.

I consider coming back with my wetsuit to experience this alluring water at the otter's level, to look up at the fern fronds, mosses and hazel catkins from below, from outside the human gaze. I could so easily sneak in, don flippers and float downstream, pushing through the reeds and catching glimpses of silvery fish, but I can't bring myself to intrude.

Instead, I make my way back to the station and climb on a train to St Ives. Face pink from the sun and legs stiff from walking, I lean happily into the cushioned seat and gaze out of the window. If you stay on the train past Hayle and Lelant and look out as it swings along St Ives Bay, nothing prepares you for the colour of the river as it enters the sea. The train leaves behind the muddy estuary, and suddenly there it is: the cerulean blue mingling into turquoise and the white lacing of waves, and the roaring sparkle of the Atlantic. It has a gleam of blue-green magic which imprints itself on the retina, as if the sky has fallen to earth and poured brilliance all over the sand. Further out, the gradations of blue deepen, lit with mosaics of molten light. On the far horizon, the sea's indigo gathers into an undulating line. It reminds me of the splashed seascapes painted by the artist Kurt Jackson. In his paintings remnants of the wild still pulse through the skin of the fields and coast, hinting at what was there before. His water looks more like the sea than the real sea, it's bashed with vibrant white that is almost too bright to look at, as if it's been flicked from the sky onto the canvas.

Beyond the fishing boats and workers' cottages, beyond the old granite quays and crumbling industrial wastelands, everything changes. A tern flutters on the wind like a blown butterfly. The air, charged with the tremendous energy of the ocean, pours in from an open window. I catch the call of an oystercatcher, the mewing of a curlew; the land's edge voiced by wind and water; the sea-glass swirl of water that is as cold as it is magnetic.

Where the Hayle River has cut itself a deep pathway to mix with the sea, nobody dares to swim. Even seals don't venture into the shifting currents of this channel; if they do they risk being trapped in the estuary by the retreating tide. Fishermen glide down here in their old boats. For them, as for otters, the coastline is a familiar edge. They know it as intimately as the seaweeds, plants, invertebrates, birds and mammals know it. Their fishing boats are designed and shaped to be harmonious with the elements, to launch, float and balance like seals on the waves.

The roads that skirt the estuary take people and cars away towards shopping malls and eateries. But alongside it all, the otter sleeks in and out of the water, travelling upstream and downstream nightly, and most of us have no idea it is even there. When I talk to a couple who have been walking their dog along the riverbank they confidently state there are no otters in the area. 'There can't be any,' they affirm. 'We put food out for them,' they added, 'but it was never touched.'

It feels right that there are otters here, hidden somewhere in this realm. It's as if we are closely tied, but only dimly aware of the other. In this way we rest, on the peripheries of one another's imagination.

A few days later, where the river pours into the estuary, belly to the earth, I'm in full camouflage, crawling dune-ward, in a small curve of marsh that smells deeply of the river's soggy undersides. Tempted outside by the falling dusk, I've chosen the limpid air and unpredictable contours of otter-watching over the desiccating warmth of the heated living room where I've been staying.

Streetlight is falling on the moving water, painting it with sodium-orange reflections. I can use this borrowed light. Any movement, any ripple or bow wave, shows up in clear lucent lines. I've been here for three consecutive nights, following the smell of fresh spraint and the oil-illuminations of dark, glossy water. I've been listening, listening and waiting. Upriver I've found there is less potential; an otter in the water would dissolve into the reeds or glide downstream and be gone in a second. Here I have a wider view, and my gaze is summoned by the slick, reflecting surface of high tide.

Sound carries any sort of call a good distance over the water. The otter's whistle is designed like the whistle of the kingfisher, both voices adapted to travel rapidly over shifting surfaces. I turn my eyes to the fringes of the waterline, scanning with my binoculars. Seawater has flooded into this space and rinsed it with a flat metallic sheen, where the river loses itself in absolute reflections of darkening sky.

Just as the moon rises I hear it. A continual calling in a high, loud, sporadic whistle. Perhaps not a whistle. A fluting, peeping pipe-note. There is no reply. The magnification of my binoculars means that I can use the last fragments of light that remain on the water surface. And there is a slippery form, pointing upward, champing at some small marine creature. I watch as if she is the only otter in England and there is nothing in the world but me, this otter and this moment. The moon dims behind a cloud and all I can hear is a faint crunching, a licking of water, a distant car as it curls around the bay.

She dives and I catch the curve of her, and then nothing. She's down in her other-world, underwater, foraging, turning over stones in the dark. Perhaps she is a ripple-wake heading out to the banks of the stream. Watching for the otter's return alters my perceptions. I am flattened to water level, but my senses only pick up its strange, fractured illuminations, its dull silt and its reticence. Other senses begin to take over as my powers of sight diminish. I can hear the river pulling itself stickily into the estuary. The stagnant hulk of a boat protrudes out of the silt, its black oak ribs rooted in weed and barnacles. There is a hint of moisture on the moving air, a swish of rushes, water percolating through pebbles.

Stalking at night leaves you open to 360 degrees of weather, and soon I'm as soaked from the ground as from the relentless, mizzling Cornish rain. When I stand up it's so dark I can hardly see my feet and I have to rely on the faint glow of a few street lights reflecting in the water. I use my other senses to navigate the shifting ground of the estuary, and feel my way back as the fleeting searchlight of Godrevy writes a frail line of light across the black sky. Like the otter, now you see it, now you don't. What you see and hear at these edges depends so much upon where you stand. Here where the edge-land blurs into another world, its tissue seeping under your feet, it's possible to lose yourself completely.

Stream

Otter returns now, almost unnoticed,
beds down in holts, roots of trees or man-made,
reclaims its own liquid world – all the fast
sunlit currents, slow meanders in shade,
redundant mill races, still pools, misplaced
rivulets – home for a species mislaid.

Paul Hyland, 'Otter Returns'

It's June and close to the solstice; my otter odyssey has lasted a whole year. At the start I thought that perhaps my fascination would dwindle, that it might get lost. The opposite has happened. I've been bewitched, and the spell has spilled over into all the corners of my life. I've found more threads to follow than I could have imagined. I've discovered that otters break all the rules we make for them. I've begun to suspect that they vary in habits and behaviour from region to region, as if they are distinct tribes affected by local geology, river systems, light and weather. And throughout the land I've found people as passionate and fixated as myself, working to protect this beguiling animal.

With the days longer and light lingering late into the evening, I'm planning a journey to West Wales, to a pioneering farm I visited in my teens, where I hope wild otters have been breeding. Otters have re-established themselves well in this area and are known to be widespread in the local river. I hope to find some breeding sites and possibly see cubs. I may be lucky, because this is no ordinary farm.

I've borrowed a small car, and in the boot I've stashed a tent, a sleeping bag, some field guides and some rudimentary cooking equipment. The farm I knew still had cows to milk and byres full of hay and straw, but its owners, two friends of mine, had a vision for its future. Twenty years on, the domestic animals have gone and the vision has been achieved. It's now a fully fledged conservation centre.

The difficulty is, breeding otters are even more of a challenge to see than solitary ones. When a female otter is ready to give birth she may appear to vanish. In areas where there is little scrub or tree growth, she may sidle quietly into reed beds, or sometimes she may leave the main river and travel up a quieter, more intimate tributary. A pregnant bitch otter may even go into the marshy hinterland, some distance away from what we humans would think of as otter habitat. When it is time to give birth, her instincts urge her to hide away. She could burrow into a twiggy bank of gorse or reeds, or push her way into cavities amongst twisted roots at the edge of a stream or a strip of woodland. She might move into the pungent earth of a fox or a disused rabbit warren. In some places she may excavate, especially if there is soft earth or peat that can be moved to accommodate cubs. Her immensely powerful musculature means that she can dig, but she can also climb well and may claw her way up into broken or pollarded riverside trees to find a suitable hollow; anywhere that is safe and out of sight.

What we know about otters and their cubs has been observed in captive otters, or cubs that were found in the wild and rescued. Recently, the wildlife film-maker Charlie Hamilton-James filmed a family of wild otters inside a holt, but these youngsters were already several months old. In an effort to discover if otters were still breeding in the same places that Henry Williamson had depicted in Tarka's North Devon, Charlie placed a night-vision camera inside a hollow tree where he had discovered otters breeding. The result was a thrilling and intimate bird's-eye view of three young otters and their mother inside their holt, and it showed just how accurate some of Williamson's observations on the behaviour of his fictional mother and cubs had been. Otters have always been stealthy and secretive, but thirty years after otter hunting had ceased, the otter's nervous behaviour indicated that she instinctively remembered the dangers that humans once presented.

In West Wales, even after hunting ceased, only 28 per cent of a thousand sites were found to have signs of otters in 1978. By the mid 1980s the number of sites with positive recordings of otter presence had only risen marginally. Even this watery westerly stronghold had been decimated.

By the early 1990s the cleaning-up of rivers began to take effect. The toxic chemicals used by agriculture that had so punished the ecosystem had been taken out of use long enough for water quality to improve. The situation was carefully monitored by the Environment Agency and the otter's habitats were allowed to recover. Habitat and clean water were the key. The fourth national otter survey was much more encouraging, and in 2002 the number of sites with positive recordings of otters had risen in some areas to closer to 70 per cent, although there were still huge regional variations. In parts of Wales the number of positive recordings was up to 72 per cent.

Twenty years ago, in a corner of West Wales, close to a tributary of the Afon Teifi (the Welsh name for the river Teifi) where otters had once been plentiful, two friends of mine, Barbara and Neil, bought themselves a damp forty acres of land. Denmark Farm was at that time a working Welsh hill farm. What attracted my interest was their idea of digging a small lake to help restore the land's water habitats. If havens like this were created, otters could arrive from nearby, strengthen their population and gradually, habitat permitting, recolonise. Denmark Farm would offer a potential sanctuary for the otters, and eventually they would edge back into the waterways of the human environment. A breeding holt would be built, inviting otters to come here to have their cubs.

When I first visited Neil and Barbara's farm in the early Nineties, I saw that their vision extended beyond the lake. They hoped to phase out the traditional farm and create the right conditions for what was a very depleted ecosystem to regenerate. The otters' reappearance would be good news, indicating that the whole web, across the entire aquatic ecosystem, was reconnecting. Otters need clean water, but they also need sheltered places to sleep. They need trees and undergrowth and healthy fish; they need eels, crayfish, invertebrates, insects and rich plant life. For them to settle, the whole food chain must be working.

For a time the lake seemed like an impossible dream. The farm's natural habitats, and those all around, had been seriously degraded by years of intensive farming. A few skeletal trees marked the farm's outer boundaries. Ditches drained the fields and led the water away, down to a meagre stream just beyond the boundary of the land, and this in turn trickled downward until it eventually ran into the Teifi valley. To make the land more productive for humans, water had been controlled, marginalised and drained away; the marshy weeds had been kept to a minimum, and the intricate and diverse balance of wildlife that had once lived here had all but disappeared.

I remember standing in the farmyard by the cow byre on my first visit (having found no sign of any otters) and Barbara, the quiet powerhouse of many of the ideas for the project, telling me that they had moles in some of the fields and the advice was to eradicate them. The moles worked hard in those fields, eating inverterbrates and turning the soil. They were here first. The farm was about sharing earth, not poisoning it. Barbara imagined those moles and their work, and the name for the Trust was born: Shared Earth. It would be an exemplary study centre, inspired by those industrious moles.

An enormous amount of work followed. The drains were broken up, the lake was built and unpolluted water soaked back into the landscape. The guiding principle of the Shared Earth Trust was that nothing, apart from the trees, would be reintroduced; all the species that restored themselves would do so naturally. Quite soon, delicate marsh-loving plants sprang up, as water dominated the fields once more. Insects came in their thousands. Toads dug their way into the moist earth. Otters surely would be inquisitive enough to investigate? Around the edge of the lake, willows were planted, and their roots held the banks firm. One day this would be prime otter habitat. Neighbouring farmers watched what they saw as the ruination of profitable land and eyed the general spread of disorder and weeds with horror.

Friends who sympathised with the Trust sent generous funds, and many joined in, arriving in wellies to share the planting of shelter belts of trees. Hazel, hawthorn, wild service and guelder rose, ash, birch and wild cherry were planted around the boundaries of the forty acres. Blackberries and rosehips returned. The following year, on the razed site of an old bluebell wood, a mass of delicate nodding buds emerged, and the new oak saplings that Neil, Barbara and friends had planted sprouted upward through a pool of the bluest of blues.

In one of the most marshy fields, fed by tiny springs, the farm lake bloomed with a curdle of frogspawn. In that first year, water-loving seeds fell from the wind and the feet of birds. Those that had been long buried sprouted after years of dormancy. An isolated island, constructed in the middle of the lake, was entirely covered in a forest of new growth. There, a carefully designed man-made otter holt was built, using logs and branches to form hidden chambers that an intensely curious otter might like to explore. Undergrowth hid entrances and exits, providing the added privacy an otter would appreciate. I spent my Christmases on the farm, dreaming of lutrine presence, but still I never saw an otter.

When Neil told me he had finally seen an otter on their land, he described it feeding on amphibians in the frozen winter snow in the wet pasture, a surprising distance from the lake. Otter Project officer

Geoff Liles came to investigate the holt. All the signs showed that under the tangle of brambles, willow and alder, the otter had finally sniffed out a warm, dry place to sleep. Spraint was found in four of the six chambers of the holt, and a freshly built couch that an otter had made out of rushes and grass. Geoff said that the otter's recent regular visits to the lake indicated an abundance of frogs and toads, confirming that the ecological system in the wetland areas was working. 'The frogs and toads are there because the habitat is right for them,' he wrote, 'with open water, streams and wet meadows, including dense rush clumps where they can hibernate. So it's the network of habitats – not just the lake – and the fact that the food chain is now working in these habitats, which attracts the otter.'

With news of the sighting, I dashed back to the farm. I remember watching the lake holt, and as I watched, many other details of the landscape came into focus. Staring out over a valley of spider-webbed grasses and shivering seed heads one morning, with no otters in sight, I went into the meadow. A barn owl perched on a fence post turned its face to me, lifted itself silently and flapped away. My attention shifted to the frozen thistles that were standing in the meadow. Their complex concentric form seemed astonishing and the delicate spined leaves of the spear thistle especially beautiful in design. The stems of these plants are spined and winged, and the woolly down and purple heads have a gem-like, defiant aura. Ted Hughes describes the spiky form of this plant, and the revengeful way it bursts forth, in his poem about thistles. He describes the way each one spreads like a plume of blood from underground. Hughes recognised the thistle's significance

211

in the layers of our shared history, and saw in it a metaphor for our battles with the wild.

Neil and Barbara eventually moved on from their nature reserve, but it had been a huge success and proved that, with care and attention, wild nature can regenerate and restore the balance that was once there. Now the Trust lives on in new hands, and with the news that otters are doing better than ever in this region, I can't wait to get back and see what it is like, and find out how the otters are doing there.

Unable to imagine the place without its originators, I telephone Neil. Would he come from the coast where he lives and meet up? He's not sure, but I think he will. I want him to help me find otters; I feel like it's his patch and I suspect he won't turn down the opportunity of a bit of tracking.

Soon enough I'm driving north and then westward, to the huge coastline which melds into the borderlands of the river Severn. *Mor Hafren*, the Welsh sign tells me: the Severn Sea. From one side, it seems to be nothing more than the brown Bristol Channel, the baggy and frayed old mouth of what must be many, many rivers. But in its silty waters it carries sediment and deposits alluvium that nourishes the whole area. Beyond the tidal plain, the new horizon reveals the dark shapes of distant mountains where the formidable river, the longest in Britain, is born. North of the Black Mountains, on the slopes of Mount Plynlimon or Pen Pumlumon Fawr, the Severn begins, alongside its two powerful sisters, the westerly flowing Ystwyth and the southerly meandering Wye.

With its massive body of water and its many tributaries, the Severn carries more than just its physical weight. It reaches deeply into the history of how people and rivers have shaped the landscape. It was the major arterial system that fuelled the Industrial Revolution in England. But long before this, the daunting, shape-shifting Severn had already created its own contours, with boundaries and marshlands where more than a thousand years ago the advancing Saxons must have looked out over the swirling currents and treacherous mudflats at an apparently vast and unconquerable water-land. In the distance they might have seen the shape of a land that was misty, unreadable and out of reach. They called it *Wealas*, which was their term for foreigners, and just to cross over to reach these people and their land was a tremendous feat of endurance.

From the trickle where it is born, at a height of 2,000 feet, the 220-mile-long river hefts its ponderous presence through hills and marshes, gathering itself into this dour sweep of estuary. The Romans would have seen the uncanny surge of the tidal bore curling and rushing back upon itself, and as they watched it move powerfully inland, they must have been amazed. In Roman lore, as in Celtic, all rivers were possessed by something other-worldly; these spirits were veiled from the sight of humans but affected them in many ways. The Severn was no exception and echoes of its supernatural past are preserved in its name. The Romans called the river *Sabrina*, and this name might have been chosen because it sounded like the one used by indigenous people. The nymph Sabrina had a partner, a fearsome Roman water god fit to cope with the powerful tidal bore, and he rode high on the back of a sea horse.

I approach the Severn Estuary in my car on the M4, and all I can see at first is a dazzling brightness, an unending mudfield mirrored with cloud patterns. As the view unfolds, I see it is pocked all over with constellations of feeding waterbirds. I want to stop and look at the glassiness of it, at the gentle seeping below, at the vast space and wildness, but I'm trumped by the pressing stream of traffic around me. Looking upriver I wonder what would have happened if the plans for the tidal barrage had gone ahead. The good news for wildlife is that the monumental barrier planned to generate a more sustainable electricity supply from the tidal flow has been put on hold. The thousands of species that would have been affected, including millions of wading birds that would have had to go elsewhere to feed, are safe for now.

Crossing the bridge, I am drawn through another balancing act: the skeletal squeeze of this spectacular piece of architecture. For a moment light and sound strobe between the tightening web of struts as they fan from diagonal into vertical and back to diagonal, stretching and elongating with almost superhuman finesse. Inside this flickering cathedral of bright white space, for a few moments everything seems suspended: air is suspended over light, silt is suspended in water, sound is suspended in the cochlea of my ears. Then the world returns. The valves in my car engine propel diesel, it burns, and birds rise bodily between the sheet mirror of water and the metal of the road. I'm released, drawn west, into a new country.

Three foggy driving hours later I have almost attuned to the confusing words on the Welsh road signs, and I've noticed a habit developing. I've stopped by the roadside to look at more streams and rivers

214

than I can remember. This new compulsion slows me down on my journey but feels entirely necessary. The intricate mesh of water veins is the otter's road system. But 'road system' implies the same kind of solidity and permanence as a human road, and otters, rivers and streams have neither; they are by definition meandering and imper-manent. The waterways I'm stopping to see are shifting, fluid, guard-ed by hart's-tongue ferns and moss, shaped by the seep of eels, frogs, gravel and mud, and never the same from one moment to another.

At Lampeter I park at the Co-op and peer over a low, concrete wall where the ground drops away. And there is the stream, moving quiet-ly between the grassy banks of a ditch, with a scattering of beer cans and shopping bags. Clear, shallow and vibrant, it curves around and disappears, riffling, under the road bridge to join with the Afon Teifi.

Where human and otter territory overlap, on these overlooked edge-lands of towns and in slivers of ignored waste ground that do not seem to belong to anyone, amongst brambles and bottles, there may be safe havens for otters. Even if the area does not seem to be reserved for nature, wildlife will have moved in. There may be some danger from litter, and it may not be pretty, but it will have been ignored, and areas that are left alone like this are exactly what the otter needs.

I can't go anywhere now without peering at these areas. Before I plan a journey I look for the thin blue streaks on maps, trace them along contours to their waste grounds and sources. This stream, just outside the Co-op, is the water that pours off the land around the Shared Earth Trust. Some of it may have come from the springs that feed the lake at Denmark Farm.

I climb the wall and descend a bank into an awkward space, an urban edge which nobody owns, where no one would normally walk except in a private emergency. These are places to ignore or to tip out unwanted litter. Entering into this no-man's-land helps me get to know a new place. Just to hear each river's voice, however fleetingly, and pause in recognition. As the water moves and I stare more closely, the little stream reveals its world. Weed straggles on the glistening silver wires of a supermarket trolley half submerged in mud, and tiny fish swim around the wheels as they mill the current. Close-to, the gravelly voice of the water whispers about the spawning of sea trout, salmon and the camouflaged wriggle of the strange prickly bullhead. This last is an odd, mottled little fish that is one of the otter's favourite sources of prey.

After the Severn, the Teifi is one of the longest rivers in Wales. Its waters are once again restored and it is one of the most pristine rivers in Britain. It springs from the watery wildness of the Teifi pools, deep marshy lakes which are situated more than 400 metres up in the Cambrian Mountains of mid Wales. The waters descend steeply, flowing through upland pastures and raised mires and bogs, before meandering through meadow and pasture and being joined by the tributaries, such as this one at my feet.

The Teifi passes through rocky tree-lined banks, and a spectacular gorge, before finally flowing out into the sea in the wide and sandy Cardigan Bay estuary. Otters and salmon have made a comeback throughout the whole catchment, and the fish and wildlife in this area have all been watched over by the Countryside Council for Wales, which monitors and reports on the ecology of the river. CCW looks after the cleanliness of the waters and the connectivity of the habitats, and their use by the public. It seems that a sympathetic cultural shift has thrown up a new term: this river is now known as a 'Living Landscape'. The new phrase represents something vital. The way we look at rivers is changing. No longer just a method of transportation, or a fishing ground, or at worst a sewer, today rivers like this are protected by measurable criteria. As well as its economic value, we *love* the river; we swim and fish in it, and sail and canoe upon it. We need it to be alive. We pollute it as little as possible, and we restore and protect it. In theory, we know what we throw in, and we know what we take out.

A short scramble from the car park into a field takes me to where this almost-pure, ottery little tributary jiggles into the Teifi. I can see some young boys playing in the sun and diving into the water to cool off. A little further downstream the banks meander through fields, and parts of the river are hidden behind a patchwork of purple loosestrife, Himalayan balsam, elm and alder trees. I wade through the undergrowth and find a meander with a gravel beach. Already I can see claw-mark signs of otter activity; gravel and sand has been scraped into an odd little peak with some spraint deposited on the top. The water level is low. It has been dry summer weather for a good while, so on the pebbly beach I see a few weeks' worth of otter comings and goings. There are scrapes and pad marks, and a distinctive, fishy perfume haunts the air. I'm thrilled with the idea of these animals re-establishing themselves, and so close to Denmark Farm. *This time,*

I tell myself, I will be lucky and see an otter here. I make my way back to the Co-op to buy some camping food and supplies so that I can dig myself in for a good long period.

The Shared Earth Trust website said I could camp in the hay meadow and when – after a few miles up into the hills, through tiny meandering lanes – I get there, a wheelbarrow awaits me, in the quiet farmyard car park. Nobody is around. I pull the barrow from its station beneath a welcoming information board and look at the new illustrations which show where to go and what to see. Then I barrow my bundle past the reception and the classroom and find my way up to the camping field. My tent perched on a narrowly mown strip (a concession to people who want to sleep over), I arrange my tent flaps so that they open onto the shimmering mass of grasses and meadow flowers. These are strictly out of bounds, but I can lean out and peer into the buzzing jungle of stalks and pollen. Tiny spiders have closed off most areas with their barely visible netting of silk tripwires; a trail of ants crawls through the woody bases of the grass stalks, a tiny white moth, the colour of low cloud, settles on a leaf.

I turn over, close my eyes to the hot blue sky and listen. My ears attune quickly. There are so many different layers of sound and whirring voices, looping over one another with melodic abundance. Close-by is the cooing small talk of a pair of collared doves and further away I can hear birds conversing over what seem like vast distances. Gnats, midges, blackflies and a huge aircraft carrier of a dragonfly whizz around in a maze of aerial acrobatics. When it settles, I see the hawker dragonfly has a fearsome striped, armour-plated thorax and eyes like aeroplane cockpits. I think of the scintillating panorama it must be seeing as it hunts. My guidebook tells me it doesn't only live near water, as I had thought all dragonflies did, but actually stalks smaller insects in any tall grass, especially meadows. Here in front of me it is doing just that. It zooms low over the grass tips, a stealth hunter, perfectly adapted for its purpose. It dips, snatches a victim, crushes it and zooms away to predate in the next field.

I muddle some cheese, crackers and fruit together and eat lunch in the meadow. All around me, the grass and trees rattle with noise. A shape glides right over my head and forms into a huge bird of prey. I can see its head swivelling as it watches me. It's not a buzzard; it moves differently – its long, forked tail and lanky, raggedy wings swivel independently as it circles around the field. A red kite. Another bird I have not seen before swoops into view. Through the lens of my binoculars I can make out the black, white and scarlet livery of a great spotted woodpecker. It swoops back and forth across the meadow between the tall, clustered rows of trees that we planted as shelter belts years ago. Just then, a familiar wiry figure appears. A smile breaks out across my face. The same as he always was, Neil strides up and gives me a hug.

We share a cup of tea and catch up. Unable to pack in all that's happened in the last fifteen years, we give up and set out into the grass with a fluttering hem of butterflies around our legs. Their wings spray the grasses with powdery golds, speckles, browns and ambers. These are the ringlets, coppers and fritillaries Neil worked hard to retrieve, and the conversation switches to his work here. He explains just how many species there might be in the soil under one single human footprint. In a typical rough pasture, there could be 1,100 species of invertebrate under one size-nine boot, compared with just 90 in a cultivated ryegrass field. I try to say what an amazing thing he started, but he looks uncomfortable with this attribution of responsibility, so I drop it. Instead, we talk about the conflicts in choosing some species over others and the consequences and effects this may have on the whole field. Neil's project was to allow nature to make the decisions, without harm or loss of any wildlife, and simply to monitor the species as they restored themselves.

We look at the hay meadow and the rough meadow, and Neil tells me that although people prefer to look at the flowers in the hay meadow, the less attractive rough pasture contains a far greater diversity of species. In my notebook I try to scribble down everything he says: just one hectare of this boggy, reed-filled field can support 207 million ground invertebrates! We walk around the outer woodland of the reserve, where now the wild cherry, hazel, beech and hawthorn belts are producing a potential habitat for dormice. These gorgeous, beady-eyed little mice were thought to be absent in this area. Neil's acute eye, the same eye that obsessively mapped each species of grass, wildflower, butterfly and bird as it returned to the farm, now picks out some of the trees which have grown so well that they need thinning out. He reaches down and picks up a nut that has been nibbled in a subtly different way from usual and his eyebrows furrow and rise. Could it be dormice? We pass fox paths and badger latrines and the innumerable nest boxes that Neil put up years ago with enthusiastic groups of schoolchildren.

In the marsh field, the damp, gauzy rust colours of sorrel and rushes mingle with buttercup yellows and deep greens. The swaying blend of colourful stalks and flowers makes a dizzying mosaic of colour. It contrasts vividly with the heavily grazed green desert of ryegrass beyond the fence. Inside, amongst the sanctuary of the trees, flycatchers, great tits, warblers, robins, wrens and chaffinches bicker in a frenzy of feeding and chick-rearing. Outside the perimeter, the green monoculture is a reminder that the wildlife corridor is still fragmented.

After a while Neil heads home. Alone, I try to shed what feels like a sadness I shouldn't be feeling. Ownership of the Trust has moved on, and what is here is glowing with ebullient, undeniable, unstoppable life. It isn't about people. It's not about any of us. It's about what we invest and leave, and what is to come.

I make my pilgrimage to the lake. Everything has grown so much since I was last here that the water is entirely hidden, and for a moment I wonder if I will find it. I don't see it until I am very close; the willow, alder and birch trees are now so rich and tall they form a protective barrier all around. A narrow path has been mown into the grass, so that people can approach, and the neatness of this accentuates the ebullience of life around this wet area. If I were an otter, I would certainly come here. But try as I might, I can find no sign. I know there is a holt on the island, but the undergrowth is so well established that it's invisible. The artificial lake appears to be completely natural, and its hidden man-made holt fulfils its secret purpose perfectly. Since there is no indication that there is an otter holt in the information provided, many people would trail past oblivious that there could be an otter, or even a family of otters, curled up just metres away from them.

A brief inspection of the stream and drainage channel that run down to the tributary of the Teifi throws up more questions. It has obviously not rained for a while, and the ditch and stream are dry. Would the otters still come? The only time they have ever been seen here is in winter, but this may be because there was less cover to hide them. Do they come then to feed on amphibians when there are fewer fish to be had lower down in the river catchment? Within their territory or range they move around following food supplies, like nomadic human hunter-gatherers might have done before they tamed themselves and the land with agriculture.

I scour and creep and look. There is not a spraint or path or slide to be seen. I sit close to the water. In the sunlight, bright yellow petals of flag iris are beginning to unwrinkle themselves from their green cases. A breeze makes fluttering wings of all the leaves around me. A spotted flycatcher flits rigorously back and forth with tiny crushed things in its bill. Now I am still, I see a willow warbler perched high over me; it opens its beak to call, and I finally put a name to the fluting trill that has been falling all around. It preens its plumage and ruffles itself like a fledgling, punctuating each line of song with careful attention to its primary feathers, breast and tail.

A bright blue damselfly flits past and lands close to my face. I am near enough to see the transparent gauze of its veined wings, the black hairs on its legs, the taut body relaxing on the leaf. When another approaches, it flickers its wings in response, and when I look again the

222

wings seem almost invisible. Now that I am paying more attention, I notice that there are not two damselflies, but many. I am inside a whole herd of damselflies, and I'm baking in the sunlight as it streams down. Amidst this thicket of meadowsweet and bramble leaves, everything sweats with life, every surface glinting with a different insect. Cuckoo spit bubbles limpidly, and a crane fly blunders past my face like a soft, out-of-control motorbike. A clatter of rooks draws my attention upward, as a mobbed buzzard flaps overhead, its underwings reflected in sky-patterns down in the water.

The lake is secluded in its jungle of alder, willow, iris, reeds, water dropwort, ferns, horsetail and sedge. Midges, hoverflies and bees thread through it all, and beyond the flitting canopy of reflections, through the sensitive meniscus, is the warm, watery under-storey. I peer further, past the strange wobbling image of myself, and from below the surface comes the bitter-sweet aroma of slime and mould from the mud as it composts and digests. From inside the soup of this alien world, bizarre beings gaze. Dragonfly larvae and newts are suspended as if in molten amber; water snails and minnows feed on clouds of algae, trillions of bacteria coat everything and swarm in a viscid galaxy. From beneath the surface I am being watched by a thousand eyes.

I make for the shade of the woods. The path draws me into a meandering green tunnel, the patterned light invites my eyes to wander; a meadow brown butterfly lands on a leaf, then flutters to the forest floor where it is almost invisible. I count at least four different types of bumblebee. A group of birds seems to gather in a bickering delegation all around me. An exquisite spider twinkles like a jewel in the centre of the intricate trap of its web. With a gold under-body, long hairy legs and a face studded with mouthparts and spooky eyes too tiny to count, its improbable beauty and savagery dangle like a paradox in a mist of silk and thin air.

At the western edge of the reserve three horses are grazing close to the fence. One, a young chestnut mare with a white flash on the bridge of her nose, notices me and comes to the sound of my voice. We stare at one another over the fence. She is cautious and curious at the same time. Her sleek back catches the light as she moves closer and tilts her ears my way. I stare into the depths of the huge brown eyes, stretch out my hand and touch the soft muzzle. The warm, sensitive exhale of her breath touches my fingers as the tickle of whiskers and the velvet upper lip reach out to find contact.

A few days later I'm inside the classroom at Denmark Farm, and Rob Strachan, the otter man from the Environment Agency, has come to give a class on understanding British mammals. This is one of the Shared Earth Trust's many training courses. It's the second in a series of classes on the ecology of mammals. The first one was about smaller, nibbling things such as voles, squirrels, hedgehogs and shrews, but this one is about the hunters and predators. I've decided to join in, not because I want a certificate in field ecology, but because I know Rob is an otter specialist. He has played a leading role and made significant contributions to the many national otter surveys that have told us much of what we know about the recovery of the otter in England and Wales. Having paid the small fee, I sit down at the classroom table along with an interesting assortment of ecologists and we get ready to learn.

Rob gives everybody an A4 ring-bound journal in which to make notes and informs us we are going to be doing some fieldwork, looking for tracks and signs. I'm so excited about being in a classroom where an otter expert is the teacher that I can hardly sit still. I'm already planning how to set out my new notebook when Rob lays a tantalising array of mustelid pelts on the table before us. They range from the petite weasel to the largest and most luxuriant pelt of all, the otter. I have never handled an otter pelt and my fingers are twitching with impatience to get hold of it.

The lesson starts and I try to restrain myself as we pass the pelts round. The weasel's is smallest, with an indescribably silky surface, softer than the silkiest ginger velvet. Then comes the slightly larger and darker stoat, and the stoat's lovely white winter version, the ermine, with its distinctive black tail-tip. These were often depicted in royal portraits, with hundreds of the tiny white pelts, tails and all, neatly woven into a decorative cape draped around this or that monarch's shoulders. Rob's PowerPoint presentation shows how the stoat's fur changes with the climate; as winter snows retreat, in many places the stoat no longer needs its snowy camouflage. Likewise, we learn that the snow hare is finding itself in an awkward position. As the spring now arrives earlier in the Scottish mountains, the hare's white winter coat that once kept it hidden in the snow is not shed until long after the snow has gone, which makes it stand out, so now predators can spot the poor hares more easily.

Next up is the more mottled and less stylish polecat, whose name, Rob explains, derives from the French *poule chat*, meaning 'chicken cat', because of its liking for domestic poultry. This fur is slightly coarser. Now we are seriously multitasking, as we look at the Power-Point pictures, listen to Rob, take notes and handle the furs. Rob has said we are going to be tested later, and my brain is feeling widened and stretched in ways it is not used to. There is no let-up; we examine the superbly thick and lustrous fur of the coveted mink, which was bred for fashionable fur coats. The mink's face is pointier than I had imagined, as if it has been sharpened like a flint arrowhead. Then, at long last, I am allowed to touch the otter pelt. Huge in comparison to the others, the otter fur is heavy. It is chestnut brown, much lighter in colour than I had been expecting. Henry Williamson once described the colour of a dry otter's fur as the same shade as the exhale of a puff-ball in autumn. I stroke my hand along the thick, furry length, all the way to the tip of the tail; I lay my cheek on it and think of the original creature that inhabited this fur. This contact might be the closest I'll ever have with a wild otter. When I notice the others watching me, waiting for their turn, I reluctantly pass it on.

After the furs, we handle the skulls. I'm enthralled at the size of some of these. At just 40 mm long, the weasel skull is mouse-like and fragile, which belies the outlandish strength and ferocity of this tiny mammal. Legend has it that a man shot an eagle and when he looked at his trophy, there was something strange. There was a weasel skull, locked by the jaws into the eagle's neck. The weasel must have just done what it knew how to do. When the eagle pounced, the weasel fought back and never let go.

The badger skull, at about 120 mm, is dense and heavy. This is the sturdy brain-case of a muscular underground earth-mover. Block up a badger sett with spades and rocks, and they simply wait until the enemy has gone, and then dig their way out again. They have long claws, like small bears, and leave chunky bear paw-prints which are wider than they are long.

There are fox skulls, seal skulls and, dominating the whole collection, my favourite: a magnificent giant otter skull, a monstrous two feet in length. This otter could do some serious damage, I think, admiringly. Before Rob has time to explain that it is in fact an extra-large demonstration model, for teaching in big lecture theatres, I get my photo taken with it.

Rob shows us how to tell a young fox and badger skull from older specimens. In the juvenile animal, small cracks or cranial fissures are still visible, where the skull has not yet undergone full ossification. The older badger has a more pronounced sagittal crest, where the fused bones have formed a ridge which increases with age in the adult animal. We learn how the badger's strong wrap-around jaw does not disarticulate – it doesn't come apart at the joint like other jaw bones might – so the badger has one of the most powerful bites of many mammals. I remember a friend's dog, too curious for its own good, sticking its head down a badger sett and emerging with horrifying puncture marks, streaming with blood. The dog survived but must have been haunted for a long time by the striped snout and crushing jaws coming out of the earth at him.

I'm caught between wanting to handle and stare at these bones, and note down the fantastical dictionary of scientific language Rob uses. The weasel skull has a tiny hole in it close to the eye sockets where it picked up a very common parasite that weasels can apparently catch from eating mice and other small prey. This parasite is a nematode worm which has an unpronounceable name: *Skrjabingylus nasicola*. I can recognise in the word that there is something to do with the animal's nose from distantly remembered Latin, but that is all. While Rob speaks, I have the tantalising feeling of glimpsing a chink of light through a door to a monumental library of precise and intricate knowledge.

Rob reaches superhero status when he finishes his stunning slide show with a picture of himself kayaking with a pod of orca, and finally, most impressive of all, waving from inside a shark cage which is being lowered into a choppy ocean. Applause breaks out, and after a brief break for coffee I find this is not the end. We settle down again and he talks us through the history of the conservation legislation for the protected species. By the next coffee time we have covered so much that I feel as if my head is going to burst.

In the morning we start our classroom odyssey once more, and I fail to pace myself, writing so many notes that my whole arm aches. In the afternoon we go out into the field, making a wet foray into the hills to find signs of pine marten. I follow close in Rob's nimble tracks. He is totally waterproofed, perfectly adapted to scrambling through bogs, up gorges and over streams. The mist is falling, but we carry on regardless and I can't stop listening, lulled by the sound of Rob's encyclopedic tones. I want to know everything he knows. Here in the outdoors, Rob comes into his own. Now he is a teacher of the most ancient sort: a tracker, crouched down, sniffing, gazing into the grass, gathering information, learning in the same way as our ancestors. Without fuss or nonsense he is carefully, quietly, passing his knowledge on. I'm reminded of the American animal tracker Tom Brown Jr. As a child, Tom had been fascinated by nature, but a chance meeting with an ancient Apache tracker named Stalking Wolf gave him some tools, as he put it, to 'track the mystery to its source'. This idea, that one can learn to put oneself into the skin of an animal and intuitively find where it went and why, is the traditional tracking that was performed by our most ancient ancestors, those who depended on the tracks of an animal for their survival.

'The first track is the end of a string,' Tom Brown writes. 'At the far end a being is moving; a mystery, dropping a hint about itself every so many feet, telling you more about itself until you can almost see it, even before you come to it.' This sort of tracking involves an intense level of sensitivity to the ground and the stories it holds. Rob has had a long scientific training, but there is another dimension at work here; he also shares the skills and senses of a traditional tracker.

When a group of us walk down to the Teifi to do some otter fieldwork together, we begin by moving quietly towards the river, walking through a field full of grasses and clover towards some steep banks. Rob shows us how the landscape has been altered, where the riverbanks have been built up with rocks fixed firmly in baskets of strong wire. It looks like they have been put here to prevent erosion, which sounds good. But, as Rob points out, this is engineering designed for humans, not rivers. It is here to keep the river flowing along one fixed course, which being a river it would not naturally do.

We look at where the river has not been altered, where the banks tumble downward to the water and a patchwork of natural habitats thrives. There are reed beds, patches of marsh, sandbanks, fronds of weed, water meadow, gravel beaches, rocks and root-strewn cliffs. The water works its way around all, crammed with fishes and a whole array of invertebrates and water life. In places I catch the scent of meadowsweet, invasive Himalayan balsam and, in the spaces between, purple loosestrife pushes spectacular spires skyward. Patches of scrub and small trees create shade and shelter, shale lies in banks and, in the shallows, weed and watercress grow in harmony like a wild, watery herb garden.

I'm not used to tracking otters with a group of people, but these people do not trample and chat, nor do they ruin it all. They are ecologists; they move quietly and they wear clothes that don't rustle. Quite soon, one by one they disappear completely and I don't spot them again until over an hour later, when they are still absorbed in hunting about in thickets or sniffing along the rough edges of the river beaches. Gareth finds an otter sandcastle and some fresh spraint. Rob is wriggling stealthily through a copse of sallow and elm trees on the riverbank, Alison has her nose down where there is some rare plant life. Vaughan, who is mainly a hare and water-vole man, points to some spraint that is so camouflaged I can't see it, even though it is right under my nose.

The river is obviously loved and used by local people for swimming, canoeing and fishing, but this human activity doesn't seem to have frightened off the otters. There are plenty of signs of them. They must slip out at night, when all the disturbance has gone. We can see where the otter came out of the water, its sinuous pathway through the undergrowth cutting off a bend in the river. We collect the spraint, seal it in tiny transparent sachets and take it back to the classroom for analysis.

Staring through my microscope at the collection of outlandish pieces that have passed through the otter's gut, I learn to distinguish salmonid vertebrae from stickleback, and thoracic cyprinid vertebrae from caudal, and frog bones from eel gills. All those bits I'd been finding now suddenly had meaning. We find beetle wing-cases, tiny jawbones and even minute teeth, and some fur, probably from a small vole. When on a new slide I find a pair of tiny eyes staring back at me, I put my hand up. After peering through my microscope at the mysterious eyes, Rob explains they might be from a dragonfly larva, or even an adult dragonfly. I sketch the eyes, gills, vertebrae, ribs, wings and jaws in my notebook and we list and label all the prey our otters have digested. What the microscope doesn't show, Rob explains, is the soft tissue that has been eaten by the otter. Invertebrates such as slugs and worms may form part of the otter's diet but they do not show up. Also, the flesh of larger fish like adult salmon will not appear, nor the bones, because they have simply been too big and crunchy to swallow.

The poo only shows part of the picture; other parts of the puzzle, including the many night-time activities of the otter, remain a mystery.

When the lesson ends I ask Rob about the otter holt on the lake. It turns out that he knew nothing about it. With the day becoming a blur, I retreat to my hayfield. The red kite sails right over me and the woodpecker swoops repeatedly back and forth to feed its young. Neil first told me about this hay meadow and the rough pasture, and the differences between them. He showed me the patterns of what grows where, and how each field's cycles and intricacies are extraordinary; how the rough meadow supports far more species, but people love the hay meadow for the flowers and grasses. What strikes me above all is how much knowledge there is to share and how generously people want to share it. Neil knew, and still knows, this land in all its wrinkles and dips and phases; he still has an intimacy with it, a connection as great as one might have with an old friend. I wonder how it's possible to turn away, as he did, and leave the land for some other place which does not hold us in the same way? Where a place has been home, has caressed our senses in a familiar embrace, carried us with its contours, nurtured us in a mutual and unspoken understanding, how do we deal with its loss?

When I look up to the call of an unfamiliar bird on the summit of an ash tree, the air is pale and the sky fading. I pick up the binoculars and focus on the slim brown shape and subtly blended plumage, its beak gaping open and dark throat flowing with warbling song. I have never paid such attention to the tune of a bird. As the syrinx vibrates and waves of sound emanate from the wide-open bill I can feel the sonority moving through my ribcage. Now I have heard it so close, I may recognise this bird again in the morning as its solo bursts and weaves out of the trees, but I'll still not understand what it means.

In the trails of dew at dawn I find marks which suggest that an otter has at last visited the lake. It has been and gone with scarcely a ripple. There may have been a ream in the water, a subtle wake. This is a creature that melds into its landscape, leaving little sign of itself. I know I will not see this one. I could stay and watch and wait but it would make itself invisible to me. Whether I see it or not, holding it in my imagination requires something of me that is greater than all the weight of experience and scientific knowledge that I've been accumulating here. This something I have noticed in some people who work with the land. It's about a different kind of sympathy, one that is so light and unselfish that it can detach itself easily. I look at the track of the otter and know that its life will be short, like every bird, insect and leaf that dwells here. Like all things, it will eventually let go and give itself wholeheartedly back to the earth. My senses kindled, my two feet rooted, I can only ever bear witness to this.

August. The season leans towards its edge, and down at my local river the seed heads droop heavy with exhaustion. Teazels, docks, hogweed and old water mint rot into a pungent compost. Skeins of geese haunt the sky each evening, the fluting rhythm of their wings creates a soft requiem for the summer.

When an otter dies, it disappears into earth or river leaving no obvious sign. Even with the attention given to the otter, with meticulous improvements to its habitat and generally cleaner water, a wild otter's lifespan might stretch to only three or four years. Many must die through scarcity of food, and in the winter the pressure on the food chain increases drastically. Otters have to feed for far longer in the winter, and they may need to travel further. At this time, seeing them becomes easier, as the landscape is worn so thin it can no longer hide everything. I want to go back to Neil's lake to see if I can find that otter, and I travel back as the season tumbles into sudden cold. In late October, in the early morning and late into the dusk, I patrol stealthily around Denmark Farm. But there is still no sighting.

One morning I can see that the lake has begun to freeze at the edges and the undulating land is like a great bear going into hibernation. I remember years before, when the inside of Neil and Barbara's front room was a cave full of firelight. A huge and aromatic Christmas tree had been decorated with candles. On Christmas Eve Neil had prepared a vat of steaming blood-coloured Polish borscht, and its odours mixed mouth-wateringly with the pine-scented tree, candles and woodsmoke. We had planted trees in this very field all day, before coming in to candlelight and glistening soup. In the flickering light the soup tasted of the earth we'd been working with, and the bedrock, and all the minerals that flooded and washed from it seemed to nourish our blood. Those days are long gone, and much has changed, but now once again I tread through the meadow of sagging thistles. It's November, and everything is weighted under a twinkling glassiness of frozen crystals, and this time I am entirely alone. Layered with thick socks inside my wellies I crisp and crackle my way through the ice meadow of plant sculptures and down to watch mist moving over the white lake. It's an early cold snap and West Wales is blanketed in ice. All life has retracted into itself, and I wonder how anything wild can survive.

For seven evenings I walk around the boundary of the fields. I go in all weathers. I sit still. I wear the same old clothes, those that don't make a sound or give off that human crackle that must sound and smell like fire to anything non-human. I go in different lights; in the glow of sunset, in low cloud, with the stinging patter of snowfall on my face. Each time, although well covered, my fingers and toes are bitten with cold.

On my very last evening, as the light is fading, just like Neil described to me years before, I hear something which makes my hair prickle. A strange crunching sound is coming from the direction of one of the shallow field scrapes. Nothing is visible that could be making a sound like this. I move slowly towards it, blood thrilling in my veins. At last I see a thin form, close to the icy puddle of the scrape. Its humped back and frenzied movement are unmistakable. An otter. Then, its body curls unexpectedly and what I thought was one animal morphs into two. It's a mother and cub!

The sound of them, magnified in the freezing air, is coming from their front claws as they scrape and dig fiercely at the frozen clump of rushes. I walk quietly closer, until I am only a few feet from where they are. The cub is slightly smaller, but seems like a young male, with a wide face, thick neck and prominent ears. He is oblivious of me. He has only one thing in mind, and using every ounce of strength in his muscular upper body he digs away at the ice, copying his mother in her frenzy to find food. Sensing the slow pulse of something hidden in hibernation buried in that clump, they both tear at the rushes. Every so often one of them stops, and munches rapidly at an unidentifiable creature. Then they move to the next clump, and carry on with the search, the cub following in the footprints of its mother. She is teaching him. This is how he will feed himself next winter, when he is alone. As I watch, snow begins to fall and subtle flakes land on the otters' dark outlines. As each flake settles, it stays for only an instant and then quickly melts on their fur.

Whenever we encounter extraordinary wild creatures, it takes a few moments to adjust. Our senses register a strangeness for a split second. Then we might feel shock, as a prickle of recognition goes through our body. The sensation is redoubled when we can name this experience as a living collection of fur and sinew. It is *fox*, *otter*, *badger* or *hare*. The alien movement of a wild animal is like nothing we have seen in pictures or screens, and perhaps at these moments of recognition we are at our most alert. Encroaching more and more into the wildscapes of these animals, we have been forced into new contact with them, and they can remind us, by their appearance, that the divide between urban and rural, wild and civilised, between us and them, is not what we might think.

I stand in the cold, my lungs empty, my feet numb, cold swimming up my legs from the permafrost. I watch the otters thread through the lap of the field; they are thin as the edge of a knife, but powerful in their wiriness, using all their strength to break up the ice with jaws and front feet. The forepaws are extraordinary. How could I not have noticed the length of otters' toes before? They are like long, slender fingers, but with webs. As the mother otter burrows into the frozen tussocks and chews with her teeth on the ice I can almost taste ice in my mouth and feel a numbing fatigue in my own arms and hands. As I watch the otter's arms working, I sense the gnawing hunger, the force of adrenalin, the sharpened determination to survive the winter.

I speak, and the young otter turns its head. He looks me over and for a brief moment our eyes lock. Then the enchantment breaks, and the otter returns to his feeding. The mother otter simply ignores me. What impression they have of my whispering figure I can't know. What does an otter think? They may have lain watching me many times, invisible in the ditch, brown shadows, waiting for the intrusion to disperse.

I can't remember what I said, and I'm glad nobody human was there to hear, but for a few moments I understood that once, before we forgot how, we followed the furred and feathered creatures around us. We listened to them and watched them. Indigenous people remember this, and still tell stories that explain how some animals came into being: they say the wild creatures were humans who a long time back had chosen or been turned into their animal shapes. If you look at it like that, in spite of the harsh life these animals lead, in some way the covetousness we might feel for animals' skills and resilience can be explained.

These days we may only ever have any sort of kinship with our domestic animals, with our pet dog or cat; or with our chickens, sheep, horses or cows, or the wild birds that we find through the lens of our binoculars. They can become like garden animals, that we like to feed with nuts and seed until they are semi-tame, ornamental things. The instinct that drives us to collect and own is only one of the forces in our relationship with animals, but it wasn't always like that and somewhere deep in our DNA our awe and wonder remain. I feel it each time I encounter the wildness of the otter in its environment, and here in this freezing field I sense it again. I stay until the otters choose to leave. I watch their thin curves slip together into the twilight, like fierce, muscled ribbons, darkening into the ribbon of the stream.

Hunting Ground

This landscape of shadowed voices, these feathered bodies and antlers and tumbling streams — these breathing shapes are our family, the beings with whom we are engaged, with whom we struggle and suffer and celebrate.

David Abram, *The Spell of the Sensuous*

In November the rain and sleet begin, and continue for what seems like weeks. At home in Devon I listen to reports of extreme conditions; floods and early snowstorms are blasting the country from the north, and the news makes my heart beat fast. Rivers are swelling, and coupled with this, cold is tightening its grasp on the land. What do otters do, in this weather, when there is little to eat and their element turns to ice?

On the edge of starvation, they have to move. Flocks of redwing have already arrived from Iceland and Scandinavia, and in the north, snow bunting and the long-haul whooper swans are tracing their flight paths. Millions of arctic birds migrate south each year, travelling from the far north along a route called the East Atlantic Flyway. They fly thousands of miles from their arctic breeding grounds, heading to Western Europe and Africa. Many of them are waterbirds. I'd like to see the geese arriving. Across vast distances, following the tilt of the earth, flocks of brent, barnacle, Canada and pink-footed geese come. I'd like to hear the song of their flight with its urgent, heartrending soundtrack, like a chant that makes the nerves tingle.

My ancestors migrated from the north, possibly to avoid the ravages of the winter, but I feel drawn there when the weather patterns are shifting, as if by some memory in my bones. And now there is even more reason to go, in fact my research insists upon it. Otters can't change their coat in winter like the stoat or the snow hare, nor can they spread their wings and cross continents, so what do they do to survive?

The weather is relentless. Last year, near my mother's 400-year-old house in Cumbria, the river Cocker burst its banks. The townspeople there, shocked by their normally tame river turning wild, watched their livelihoods and their homes being chewed up by surging waters. Now it looks like the rain is threatening to cause the same catastrophe all over again. Alongside my concern for all those affected, I have extra worries. Rivers in spate quickly become inhospitable to wildlife; otters can't fish in stirred-up torrents. The 2010 national otter survey reported a good population recovering in the North. Do otters instinctively understand what to do when the weather begins to turn? The survey reported that in Cumbria and Northumberland where river habitats have been cleaned and improved, otters are widespread once more. But they are still vulnerable to the whims of the climate. Potentially, hundreds of otters could be affected and their fragile return undermined.

As the floods give way to frosts and ice, I can't resist the urge and I become a nomad once more. I'm following hazardous predictions: the forecast for high ground is strong wind and blizzarding snow. My journey thrills with cold. I know that otter tracks will be easier to find on snowy ground. With my survival equipment tucked in a tiny hired car I pore over the wriggling contours and sinuous blue lines of my OS maps. I decide to travel through Cumbria, do some tracking in the south of the county and stay in my mother's house further north; she has taken refuge in the milder south, leaving her home blissfully empty.

Following a tip-off from James, I begin with an important stop-off in the southern Lakes. James has his finger on the button in the otter world. He's sending me to see his friend John McMinn, who lives in a place where signs of otters are regularly found. Before I get to John's house I pass a tidal estuary where water meanders and fans out into the sea in a wide swirl of sand and silt. This kind of soft landscape is great for finding tracks, but in bad conditions success is not guaranteed. Water in this landscape is continually on the move; it drops hundreds of metres from where it falls in the mountains, and in spate heads quickly west and out into the Irish Sea. The weather would sweep every sign of otters away. Otters have been seen here, however. John apparently spends significant parts of the year tracking, surveying and photographing them, and I know he'll point me in the right direction.

It has rained for the whole journey, and I arrange to rendezvous with John in a lay-by in a hamlet miles from anywhere. 'It's easier this way,' he shouts through the crackle of my mobile phone. 'The directions are too complicated.' He pulls up a few minutes later, and we jump out into the rain and shake hands, wet trickling down our necks. We drive off in our separate cars, through tiny lanes to where John lives, high in the hills. We pause by a reservoir, and leaping out of his car, John shows me where the otters go. Even though he has been at work all day, and night is falling, the otter-talk is enthusiastic. I've found this a characteristic of all the otter people on my journey, each one more than willing to pass on information and give me advice or a helping hand.

Finally we get to his home and it's still raining. He lives in a glorious converted barn with sweeping views – even in this weather we can see far over the otters' territory – but there is no time for a cup of tea. There isn't much light left, so we put on our wellies, fleece-up and head straight out. Over the fells behind his house, the sky is baring its teeth and spitting sleet. I wonder why he's taking me uphill, but in a gap between showers I understand. He's showing me the lie of the land, giving me a feel for where otters might go. Instead of telling me things, he asks mysterious questions to see if I can puzzle out the answers. This strikes me as a very ottery thing to do, and I like it. John looks at me shrewdly. 'If you were an otter, which way would you go?' he asks, with a gentle Cumbrian lilt. 'How far do you think an otter can travel up here?' I can see the view to the south; the huge estuary of Morecambe Bay sleeks into the tidal flats, and 180 degrees further north-west, the brown silt of the Duddon Estuary unfurls into the Irish Sea.

The river Duddon enchanted William Wordsworth, but from this unprotected angle the landscape seems utterly un-enchanting. It's lost the cat-walk gleam of summer and makes no pretence at beauty. Now the sky is pelting us with sleet *and* hail and water is running into my eyes, but John's enjoying testing me, so he can demonstrate the impressive distances otters can travel. From our blustery vantage point he indicates where he has tracked his otters. It looks like they leave Duddon Sands and the lower river valley and travel for several miles upstream to where we're standing. It's *miles* away. The otters join 'becks' – as they call streams here – and follow these between the undulating fells, travelling over the high ground to join a tiny stream which then flows down into the next valley and the two reservoirs near John's house. It's impressive to be shown a whole territory in one sweep. Otters can cross the watershed and join new river catchments, bringing them into contact with other otters, which must have helped their expansion.

Having mapped the territory, John points out that the landscape offers no shelter. It is treeless and windblown, and the tops of the fells harbour only wetland grasses, springs, heather, marshes and bog, which all appear to be rapidly freezing over. But up here the wet doesn't matter to the otters. It doesn't seem to matter to John either. I, on the other hand, am longing for warmth and a cup of tea. 'The ice and cold,' John shouts through the weather, 'might deplete their food supplies, which is why they have to range so far.' I'm feeling that *we* are ranging too far but, too shy to say so, nod appreciatively. What fish do I think would be in the Duddon? It's a salmon river, but did I know these fish are not plentiful, and could be hard to catch at any time, particularly after heavy rain? Otters may go for smaller prey as it is far easier to catch, and in winter they must diversify to survive.

My stomach is growling, and *I* want salmon, preferably cooked in sauce. Finally, John and I take cover in his house, and look out through glass at the darkening contours of the sky. The time for a cup of tea has long gone, but I can smell baked potatoes, gammon, melting butter and other good things.

As dark falls and we sit in a pool of warm light around the table, John's family arrives, along with supper. They are pleased with my bottle of wine, and it's shared out in generous goblets. Having worked up an embarrassingly huge appetite I try not to gobble my food too quickly. Outside the wraparound windows, the wind and rain swirl around in the gathering dark. Otters are out there, somewhere, uncurling with the dusk and heading to their hunting grounds.

After supper, fully restored but feeling slightly woozy, I am taken to the otter nerve-centre. John's office is neatly organised, and stacked with otter equipment and books. The computer screens – several of them – are lit up with otter images. John shows me a photo of an otter trail in the snow a few yards from his home. The tracks show where the otter bounded downhill then sprang, folded its front legs beneath it and tobogganed full-tilt down the side of the fell. I feel a twinge of envy about the view from his house, with its sweeping, treeless landscape. Where I live, you can't see far for the little Devon hills, and you have to hunt for otter signs hidden in layers of dense hedge, undergrowth and woods.

John has surveyed much of Cumbria for otters, and just like James in Somerset, he seems to spend every spare moment at it. His bookcase trumps mine for otter books, and from the corner of the room, a stuffed otter stares through spookily life-like brown-glass eyes. The teeth are unnaturally bared and still sharp. It died young, hit by a car, and its preserved body is now used for education; John takes it along to show at his talks. 'They never snarl like that, though,' he assures me. 'The taxidermist did that – he said that most people want a stuffed animal to be *doing* something. And you know what, he had to do another otter the other day, and the guy who ordered it wanted it with a "wet-look", standing up on its hind legs, with a fish in its mouth.'

Having a real otter, even a dead one, John can demonstrate to people its texture, shape and bulky dimensions. People are often amazed at its size. This is because, John says, they're often misled about exactly how big otters are; many people are used to seeing Asian short-clawed otters in zoos, and they think that all otters look like this. But that species is far smaller, friendlier and cuter than our shy native Eurasian otter.

A network of local people regularly brings John news of otters. Like James, he collects corpses that have been killed on the roads. Sometimes he is brought live otter cubs that have been found abandoned or orphaned. 'It's difficult not to get attached to them', he tells me. 'Otters are like any baby mammal, they need feeding all through the night. We didn't get much sleep,' he admits. 'And yes, they do bite.'

Again I feel a twinge of envy, in spite of the teeth marks. Before they were released, John held and cared for these creatures, and this is something I will never do. I wrestle with feelings provoked by the image of these helpless, dewy-eyed bundles and a thought rears its head. For me, the otter is a symbol of wildness. What happens when it is brought into an environment which is as foreign to it as the surface of Mars would be to you or me? Rearing an orphan may help it to survive a little longer, but millions of years of evolution have programmed this animal to be a ferocious predator, not a house animal. All it wants to do is to get back in the river and away from us. All the same, I can see how you might bond with a baby otter. Their big, inquisitive eyes, helpless expression and fluffy demeanour are unavoidably alluring. Fortunately, the needle-sharp teeth serve their purpose, and can leave nasty infections, a reminder that these animals, however beautiful, are not made to be pets.

We spend the rest of the evening talking and looking at John's collection of startling otter photos. He gets out his macro lens (an impressive one) and his state-of-the-art camouflage equipment. Coupled with superhuman patience and stealth, these have helped to produce some superb close-up shots. In some of the photos, you can see how effective the otter's camouflage is; the water on the fur produces a blur effect, reflecting light and giving a strange soft focus. It's not just the otter's colour that keeps it hidden; it also uses the water and light to its advantage. This is why when an otter encounters people, John explains, it will often remain very still until they pass by. Rather than moving, which might give away its presence, the otter knows its best ploy is to stay still and blend.

And he has come across some fearsomely large-looking specimens. One picture, taken under a bridge near John's home at night, features a magnificent dog otter approaching the camera. I notice that its huge, hunched form is bear-like, and when I suggest this, John admits that he felt nervous when he saw it coming out of the water at him.

When we witness an otter, we normally see a fraction of its body. Lying on the water, with its eyes, ears and nostrils on a horizontal plane, the otter only needs to show a sliver of itself. Seeing the full bulk of its body can be a shock; an adult dog otter can weigh over eleven kilos, which is more than many badgers.

John chooses one of his pictures for me to keep. It's of a pair of otters that he saw on the west coast of Scotland. In the picture, the otters are nestled close together on a rock. The male is looking on while the female chews through a crab that he has brought her. 'He spent three hours bringing her crab after crab,' John explains. I look closer at the female otter's wincing face, and at all the crunched-up shell lying around, and I can't help thinking up a caption: 'Yes, that's lovely, but is there anything else down there, apart from crab?'

I stay over in a special suite John and his wife have created for people who want to watch otters for the weekend. Tucked cosily in my bed, I can hear wild weather outside. When I sleep, my night is a flashing ocean of images; I dream of drowning in my duvet, of swirling fur, webbed claws and mercurial bubbles.

I emerge damp and bleary, when the sky begins to lighten. John is already dressed in his suit, clean-shaven and ready to go to work at his sensible job, teaching in the local secondary school. Over a bowl of Weetabix we talk about our respective careers. Soon, he says, he is going to retire and dedicate himself entirely to otters.

Before he leaves for school, John walks me up the hill and points me to the moorland behind his home. The sky is morose, and weather stalks around us threatening yet more sleet. We shake hands, battered by the wind. John looks me in the eyes one last time. 'If you want to find otters,' he shouts through the gale, 'you have to *be* an otter.' With that he's gone, leaving me alone with the weather.

Down the valley, five giant wind turbines stand like triffids, guarding the spot where John's otters are supposed to be. I trudge into the weather, clinging to my hood as the rain starts, and quite soon I spot a strange green thing flapping on a gorse bush. As I get nearer I see that it's an item of clothing. Curious, I go to inspect and find it's a pair of waterproof trousers, hanging abandoned, as if the wind has torn them from a lost soul on the fells. I pull them off the bush and eye them up. My size! There is not a rip anywhere on them, even if they are a bit muddy from being blown about the hills. No matter, the sky is darkening and these weatherproof trousers are heaven-sent. I squeeze into the unexpected gift as the rain begins to fill my boots.

The wind takes the sound of the turbines up and away until I am close-to, and here at the feet of these extraordinary structures is a continuous hum; a strange wind-factory in the sky. Around the turbines, I see the vast open distances the otters must cover. In spite of the human interference, this is truly otter territory. Water is everywhere; in the sky, soaked in the peat, running in streams and pools all around. Anything or anyone standing around for long will have their wits blasted by the humming blades and moorland wind. The otters, however, can move over the soft map of peat with ease; weatherproof, they thread through the contours of heather and tussock, slipping into the becks, far away from any roads. Why they leave the coastal estuaries and travel upstream to get here is anyone's guess. It might be for feeding, or to hide, or even to breed. The only obvious evidence that they come here is that occasionally one gets hit by a car as it crosses over the coast road at dusk.

Otter paths avoid the human ways most of the time, but close by here, a clash is poised to happen. More wind turbines are planned. Bigger ones, right on the spot where John knows his otters have their holts. The independent survey found no evidence of otters, but on my foray I find otter spraint exactly where John predicted, alongside the stream, close to the proposed turbine site.

Further downstream I pass copses of sycamore and oak inhabited by rooks and jackdaws, and watch as a squirrel moves busily up and down a rough trunk gathering leaves. It darts about on the ground and collects mouthfuls of nest material, then carries it up and rolls it neatly into its drey in the branches. After a few miles of walking through undulating farmland and tiny roads I reach the main road and see the tide-washed coast. Here human and otter intersect again. Otters have used the sheltered silt beneath a road bridge to mark their territory. Hidden from the road, in the under-wing of a busy dual carriageway that skirts the coast, otter prints are written all over the silt. Two otters have been here. There is fresh spraint, perhaps only a few hours old, and the spraints are different; one has a stronger smell. Deep scratch-marks claw the sand as if a family of small tigers has passed through. They came out of the water, scrambled up the sandy bank and left their signs on roughly made sandcastles. The ground is marked and smoothed where they have slid repeatedly back down the river's steep sides; you can see the marks of their fur and a hint of the shapes of their bodies printed in the sand. This is where I wait. There will be a little light from the road and I can hide myself against the wall of the bridge, where it is sheltered and the otters will not be expecting me.

At dusk I settle down, layered in clothing that will make the least possible rustling. I have laid a large trout, fresh from the fishmonger, near the spot I think the otters will come. John told me they like it filleted, but I thought a fillet might disappear too fast, so had left it whole. There is very little space and if the otters come out of the river where I think they will, we will be nose to nose.

In the dank, musty underworld of the bridge, the ceiling drips and the roaring traffic sends reverberations through my bones. As time goes by and light fades, the tide moves inward from the sea. Water pushes upriver and slowly licks up the steep bank.

I sit cross-legged and still as a Buddha. I can feel the thunderous ruminations of rush hour; shock wave after shock wave buffet the air around me. I think about how we are always rushing, how in our busy lives moments of stillness like this can seem almost impossible to achieve. Below, the water is dark with promise. The current moves whisperingly, ruffled by ripples or pushed into brown wavelets by gusts of wind. I'm sheltered from the wet, but after an hour, in spite of my new waterproof trousers, I have lost contact with my hindquarters.

Some time later I am tempted to calm my rumbling stomach, and move my fingers to grasp an emergency jelly baby from my pocket.

Just as I transfer the juicy sweet up to my mouth, a subtle change in the river occurs. Something is heading upstream, followed by a ream of ripples. Slowly I put the jelly baby down, saliva still pooling on my tongue. A ripple catches a thin edge of light and folds it like molten copper. Another ream! They register as nothing more than barely decipherable dark smudges against a bigger dark, but I recognise the familiar jizz in the water. Two shadows are moving towards the bank. They disappear straight beneath my field of vision, and then reappear almost immediately, up on the bank, glistening wet, printing the sand. I can't explain it, but they shine, even in the dark. Just as moving water catches the last light in the sky, their fur does the same.

The first otter is the largest. It hasn't seen me, or picked up my scent. This is sheer luck. The wind must be taking my odour away downstream, or perhaps it is my new second-hand trousers, thoroughly permeated with the smell of the moorland. I hold my breath; at any second they could notice me and bolt. They shake a spray of water from their coats, large wet feet dripping. The texture of their fur is like wet driftwood, striated and still gleaming. The dog otter scrapes up some earth and pees, marking the ground. The female does the same. When the otters are next to one another, I can see that the second is almost as big as the first, but slimmer. I am sitting in front of two otters, and they are busily sniffing around like old racoons. They find the fish almost straight away and the larger of the two tucks into it. He must be a dog otter, with a thick neck, superb paws and bushy fur around his head. He does not want to share, and snatches the prey away, his body rippling as he turns his back on his companion. He places a paw upon the fish and rapidly rips it open, belly first. I hear bones squelching and cracking, and jaws going; wet sounds, like a ravenous cat.

The otter turns his head on one side, eyes closed as he munches. He screws up his face and his muscles squeeze his eyes shut as he eats. Otters have the densest facial muscles of any animal of this size, and this is nowhere more apparent than around the jaws when they're feeding. The bitch otter makes a yickering sound and paces around, jealous of the meal. She approaches and is ignored, then she moves in by rolling in the sand and grabbing the fish from beneath. From their behaviour I'm sure this must be a breeding pair; the male bites the female on the neck in a gesture of dominance, but allows her some fish. A few moments later he pushes her down and nips, and she squeals with indignation. When there is very little fish left, just some spine and part of the tail, they slide around in the sand, cleaning their fur, rolling together in slippery knots.

A slight change in the breeze takes my scent to them, and the female's head pops up out of the tangle. Instantly, as one, they slide back into the water. In the current they become separated and I hear their call as they surface, repeating high, urgent whistles. They drift together on the current, and look back curiously in my direction, taking turns to *huff* at me in a distinctly hostile way. The water is ruffled with their bobbing forms, and they appear and disappear as he comes to her and nuzzles and licks her face, as if searching for any remaining fish. Heads, paws, backs and tails confuse and I can no longer tell which is which. Their colours stray into the water, and fluid darkness envelops them.

The next day, the weather worsens but I drive into the mountains, heading for my mother's house. I should have paid more attention to the forecasts. The wintry fells of the north-west Lake District may be at their most spectacular, with a first dusting of snow, but this is not good otter-watching weather.

It's getting dark and is still raining when I open up the old house. I unpack rapidly and light a fire. The house has been empty for weeks, but has the familiar feel of home; it's been lived in and loved for hundreds of years. The wallpaper is peeling and the doors let in drafts, but with the fire going it warms up rapidly. Built by a Quaker in the seventeenth century, the walls are several feet thick, so they never quite lose their humidity; the old stone larder has softened boxes and packets and transformed the contents into an inedible mush. There is nothing to eat. I decide to go into Cockermouth for provisions and, while I am there, take a look at the river Cocker as it flows through the town.

The Cocker, when it is not in spate, is an old favourite of mine. Recent otter surveys nearby have shown exciting results, an expanding set of sites where otter evidence has been found. In the old days, while my family strode along the lakeshore, I would dawdle along the riverbank and (before I knew any better) push my face into holt-like gaps between tree roots. I often ended up in the water and I never found one hint of an otter. I want to scan the river now and reassure myself that this river is supporting otters.

Born from the lake at Crummock Water, the Cocker seems to be perfect otter territory. Its name is from the Celtic *Kukra*, meaning crooked, and its narrow, sinuous course is hidden from view by woods and plenty of bank-side vegetation. It meanders a gentle twelve miles through private farmland to its confluence with the Derwent at Cockermouth. Recently, a rehabilitated wild otter was released near here, in a secret location, where there would be no disturbance and more brown trout than one otter could ever need. As it winds from the basin of fell water at Crummock, through the Lorton Valley, the Cocker is augmented by a network of tributaries along the way. Much of the water that drops on the northern fells collects in the reservoir of Crummock Water. On its journey to the sea, this water is forced into a short stretch of river which has nowhere else to go but Cockermouth, where it is suddenly squeezed, joining with another squeezed body of water from Derwentwater and Bassenthwaite. So beneath the roads and bridges, between houses and around the town, lies tremendous potential for trouble. In 2009, after torrential rain, Derwentwater and Bassenthwaite remembered an ancient partnership and spilled into the underlying geology where thousands of years before they used to be one. At the market town of Keswick, hundreds of houses were inundated. The spectacular reunion of all this water was made worse by a build-up of silt in Bassenthwaite Lake, which meant less space for run-off and floodwater than there used to be. Swelled by its tributaries the Gretna and the Cocker, the Derwent was forced onward, building like a tsunami. It rampaged through Cockermouth, wrecking over 1,200 homes and destroying six bridges on its journey, finally reaching the coast at Workington, where it crashed with armfuls of debris into the Irish Sea.

I park in the marketplace and peer nervously down at the river. All that water through one tiny channel. When there is an unusual amount of rain, people aren't alone with the threat of flooding. The wildlife has to move or be drowned, and otters must head for safety into smaller tributaries. As I walk along the riverside I move as close as I dare to the water, but there is no evidence at all of otters. The river is angrily high; all signs have been washed away and going any closer to the water is either difficult or dangerous.

To what extent is this inundation stressful for otters? Perhaps they take it in their stride, and move further into the land at these times, moulding themselves to the conditions. They are adaptable, this is what they're good at. But life on land, close to humans, is not without dangers, and many otters are killed on roads when water levels rise.

The next morning brings a leaden sky. I head for the source of the river at Crummock Water, where the river overflows from the body of the lake in huge molten folds. Wave after wave bounces and booms into the narrow course of the swollen river. The water spills and spreads into the woods and fields, and in a sliver of National Trust woodland at Lanthwaite, it rises, tangling through submerged roots and undergrowth. It mixes with an undercurrent of sound, winnowing the brambles and undergrowth of the lower wood. I find piles of fallen acorns and rescue some, storing them in my pocket to plant later. From the safety of high branches, red squirrels jitter at my intrusion. When our eyes meet they freeze, still as dead wood, and only the flecks of light in their beady eyes gives them away. Watching how every instinct tells this animal to turn to stone, I can see how an otter would vanish into stillness. The squirrel is bright, but an otter is the same colour and texture as its background. Where there is plenty of cover, in brown undergrowth and brown water, they would never be noticed by the human eye. I've been thinking that I'm the stalker, forgetting that it's the otter who has evolved these subtle adaptations precisely because it is a predator and lives by stealth.

In the woods it's blustery and wet, and I can't hear or smell as usual; the landscape is fluttering, shifting and moving in ways that conceal my object. Although leafless, the trees sift the wind, and my ears swim with layers of sound. It's not just the river, but the undergrowth of the woods, the whisper and rattle of dry ferns, old honeysuckle, the tangled voices in the permutations of ivy leaves, clapping and scratching against conifer, beech, larch and oak. Then a tune bubbles over the cacophony of water and branches. Its talk threads through the trees, increasing, extraordinary in its energy and loudness. The bird's voice is distorted by the woods, rolling like a river of pebbles, tumbling towards me.

A caravan of walkers don't see me down by the water. They undulate along the track; their reds and greens and purples flash and flicker, rustling and crackling toward the drier upper edges of the wood. The water's edge is a spreading lake in a field of lakes and pools. I'm ankle-deep and again I hear the high-pitched song. I crouch, waiting for the songster to reveal itself. Again and again it rolls toward me, piercing my eardrums.

I stay by a fallen alder, my hand leaning on a green pelt of moss. Some mallards and a merganser shoot by on the water. They are nervy and don't like my intrusion. I shift my position and suddenly see it, a bright thing on the edge of my vision: the white bib of a dipper. He's singing his heart out, but I can't see his beak open. It seems he's singing from his throat, weaving his song through the waft of the icy stream.

I look back at the squirrel high on its branch, and from each shift in perspective, in different leaf-light, the texture and fur and shape change. I'm in animal territory and around me is their forest, full of their eyes. From every angle it seems that it is they who are watching me, from above, around and below.

Downriver, in the Vale of Lorton, more of the river has overbalanced and the further I go, the more it has spilled into the fields. The spate has made the ground impassable; cattle and sheep linger at the edges, and ducks relish new opportunities. The whole shift forces me uphill to the smaller becks that feed the Cocker. If I were an otter, I would take refuge here. There is shelter amongst the rocks and crannies of the hills, and the tiniest streams send their rushing water down to augment the flood below. In the village of Lorton an otter was seen by the ancient stone bridge, I'm told, but it was frightened away when the crashing water took with it the bridge and its foundations.

Morning light creeps over the grass, casting tall shadows. The meadows are shivering with seed heads, vibrating with the cold vitality of moving water. Spiders have made a million twinkling lines, in a mid-air tangle, as if catching at what little solidity is left.

The next morning the weather is even worse and I venture into town for milk. The solid roads and pavements of Cockermouth seem shaky. From the main street I can hear the river; and from the repaired bridge the water is an unreliable shade of brown and full of debris; no otter in its right mind would ever brave this. Otters can't feed in murky and very fast-flowing water, and this river is in a downright dangerous mood. It's still raining in the hills and I don't think I'll stay around to find out what will happen. My eyes and ears are charged by menace in the air, the open circuit of water a jittery reminder of our tenuous attachment to the earth.

I get my shopping and head away as townspeople scurry to work or the shops, their eyes flicking toward the rising current. Back in the house, I look at an email John sent me about a family of otters caught in a Fyke net near Lindale in South Lakeland. These fishing traps are designed to catch eels, and are illegal if not fitted with a special guard to prevent otters swimming in and becoming entangled. This was a horrifying case, where there was no otter-guard. Three cubs found their way in and one by one become trapped. The mother otter, unable to free her three young, had drowned alongside them.

Otters live fast-moving, dangerous lives and humans do not often make it any easier for them. This is a creature that, although it is acutely aware of danger, lives totally in the moment. It experiences the world through the dimensions of its senses, and to us these senses seem unimaginably acute. But they have to take their chances in a perilous world, where one small mistake can mean death. The vital alertness to the present moment that the otter needs to survive makes our busy, distracted lives look very strange. We spend a great deal of time slipping out of our bodies and away from our senses. By day-dreaming, worrying and rushing about we impatiently betray the present moment of here and now. The otter's life may seem harshly short, but it does not miss a moment, and in essence might be more purely joyful than our own.

Later, I pick up my old edition of *Tarka*, and notice the subtitle of the book: 'His Joyful Water Life and Death in the Two Rivers'. I am not the first to notice what seems like the wholehearted joy of the ot-ter. I watch the landscape outside darken and the glass reflect inward, and then I open the old book once more. The narration is soothing at first, and as absorbing as it ever was. Three-quarters of the way through, the world around me has long since dissolved and I am deep inside the wildness in the pages. I find myself reading a particularly intense passage. Tarka, now in his second year of life, is being pursued by a pack of hounds. In an effort to hide, as any otter would if an enemy is close-by, he creeps into a dark cavity amongst the roots of a waterside tree, only to meet the familiar face of one of his cubs. Tarka has inadvertently led the pack to his family. With the baying pack sensing more than one otter, their fury intensifies and they move in for the kill.

I read this part of the story again as if I had never seen it, and the events that follow contain so many elements of nightmare that I won-der if a child reader would be traumatised by it. A common method of hunting otters when the animal was hiding underground was to pound the earth above the holt and make so much noise that the terrified otter would bolt, making it easier to capture. It's clear from the vivid writing that Williamson observed this sort of hunting scene many times. The cub Tarquol, younger and less experienced than Tarka, can't stand the pounding and does exactly what the reader fears and the hunters expect. He bolts straight into the jaws of his pursuers. With the pack distracted, Tarka flees, and in just a few short lines Tarquol's fate is sealed: 'Tarquol sprang up as soon as he fell, snapping and writhing as more jaws bit on his body, crushed his head, cracked his ribs, his paws and his rudder.' As if this isn't enough to win over the reader's sympathy, Williamson presents the jubilant crowd. The contrast of beauty and brutality is characteristic of Williamson's method in many of the death scenes in the book. These paradoxes are often depicted in nature in the story, but here it is the unnatural, cold-blooded actions of the huntsmen with their hounds that are so poignant: 'Among the brilliant hawkbits – little sunflowers of the meadow – he was picked up and dropped again, trodden on and wrenched and broken, while the screaming cheers and whoops of sportsmen mingled with the growling rumble of hounds at worry.' The scenes resound with echoes of Williamson's experience, not only of hunting, but also of trench warfare. He leaves the reader in no doubt which side to take.

Since otter hunting ceased half a generation ago, we have less contact with wild otters, and they elude us more than they used to. Their deaths are invisible to us, especially without deliberate persecution.

Touching an otter's fur was fairly common in the days of hunting, especially for people in rural areas, or while fur coats were fashionable. Protected as an icon of the environment, they are now cherished, and increasingly widespread, but further out of our range than ever. Their presence haunts us, and while we reach for them with our eyes on screens, in books and in still photographs, that visceral, sensual contact has gone.

Before the otter gained legal protection in 1978 it was generally considered to be vermin and otters were regularly trapped, shot and snared. For a period of six months each year, between April and October – when rivers were less likely to be either unpleasantly cold or in spate – the hunting season was open, and the pursuit of the otter was an enjoyable pastime. Otter hunting was so popular that hounds were bred to cope with wet conditions and prolonged pursuits. They were trained to identify the otter's scent, which they were able to follow in the water. The persecution continued until it was noticed that there were nearly no otters left, and a halt was called.

In many ways we have made up for our errors, but all is not rosy. Hundreds of otters still die at our hands each year. The ones we know about are mostly the otters killed inadvertently, and the majority of those that are found are collected from roadsides around the country and sent to be investigated by the Cardiff University Otter Project. Many otters that are injured may never be found. Knocked by the bumper of a bus or lorry, they simply curl up and die in a ditch. The true figures are invisible, and may be much higher than those known about, as otters are accidentally caught and drowned in illegal fishing traps, killed by litter, poisoned by accidental leaks of pollution or become emaciated and die. At Cardiff, an otter arrived from Looe in Cornwall where it had been caught in a snare. The wire had slowly tightened around the middle of the otter's body, almost cutting it in two. It was obvious from the body that the animal had undergone a long struggle. Rumour has it that a certain number of otters are deliberately 'disappeared' each year by vengeful fishery owners. Of all the ways for an otter to die, after the Fyke net and the snare, the one at Cresswell Pool in Northumberland must be nearly the worst. A group of birdwatchers were viewing the playful antics of a young otter fishing in a lake at a wetland bird sanctuary at Cresswell. The onlookers watched as the otter swam beneath an old duvet that somebody had dumped in the water. It became entangled in the weight of the sodden duvet. Eventually a boat was launched, but it was too late. Pulled out of the water, it was found to be a male less than a year old, probably one of two cubs that had been seen playing together with their mother in the same area.

Although we celebrate the protection of the otter, and its miraculous resurgence, the number of otters that were deliberately caught and killed each year by hunting is similar to those now killed accidentally. An otter might tell us that litter and cars are as dangerous as weapons. Are we in a sort of waking sleep, that we persecute the inhabitants of the natural world without realising?

Our love–hate bond with wild creatures is complex and deep-seated, and the history makes painful reading. The Bible suggests that nature, including all animals, exists solely for the interests of man:

'... fill the earth and subdue it; and have dominion over the fish of the sea and over the birds of the air, and over every living thing that moves upon the earth' (Genesis 1:26). It is almost as if war is declared. What if we substitute 'subdue' and 'dominion' with 'care' and 'responsibility'?

During the Middle Ages, a 'King's Otterer' was employed to eliminate otters, and records show that as early as 1157 hounds were used to seek out and kill them. Because otters ate so many of the fish in their ponds, and seemed to live mostly in the water, many monasteries classified otters as fish and otters could therefore be eaten on Fridays. The monks' bellies were filled and the predators eliminated.

The death knell sounded louder when in 1566 the 'Acte for Preservation of Grayne' was passed. Many creatures, now classed as vermin, were systematically wiped out, and an especially high bounty was offered for the otter's pelt. One poet spoke up for the otter, but it seems he was in the minority. George Turberville's 1575 poem 'The Otter's Oration' was sympathetically written in the otter's voice, and pleaded for justice, claiming

> all Adam's seede
> Do daily seeke our names for to distayne,
> With slandrous blotte, for which we beastes be slayne.

In spite of this eloquent protest, the 'slandrous blotte' did not go away. Not realising that otters actively maintain the health of the aquatic ecosystem, people believed that otters were stealing all the fish. The lutracide intensified when, in 1653, an angling writer named Izaak Walton made resentful declarations against otters in his book *The Compleat Angler*: 'they destroy so much; indeed so much, that in my judgement all men that keep otter dogs ought to have pensions from the king, to encourage them to destroy the very breed of those very base otters, they do so much mischief'.

In one section of his book an otter is caught devouring a fish, and Walton and his friend declare revenge by killing the otter and skinning her, for 'the gloves of an otter are the best fortification for your hands that can be thought on against wet weather'. When the victim was caught they found she has 'lately whelp'd'. The cheery group then go in search of the cubs, and set about eliminating the whole family of otters 'merrily, and all her young ones too'.

In the nineteenth century, the sport of otter hunting was increasingly well organised. Paradoxically, during this time we learnt more about them through this regular contact. Landowners bred hounds to search out any resident otter's scent along the banks of a river. The hounds were excellent swimmers, and the hunters often waded after them on foot. Horses weren't used, and the owners knew their small pack well. Each hound might have a different set of skills, and their masters were as proud of them as any breeder might be. More recent accounts show how enjoyable it was watching the pack at work, and it was considered as much a sport as a method of controlling otters. James Williams still hauntingly remembers his own otter hounds: 'I still hear Regent's tree-trembling roar in my dreams, and often see

Deacon throwing his great skull at the sky to give full vent to his enthusiasm for a swimming hunt.'

As happens for many hunters, a huge sympathy grew out of James's knowledge and experience. His passion has led to an impressive and important body of otter research, and he has written two books and many articles about it all. His successful efforts at galvanising conservation work across the whole county of Somerset and beyond have educated many and made a huge difference to the fragile expansion of the otter population there.

There used to be many private packs of carefully bred otter hounds, such as James had in his youth, but there were also some that roved the country, travelling great distances to join and help other hunts. The exciting spectacle attracted onlookers, who gathered on bridges and riverbanks to cheer on the hounds. People were caught up in the action, even if they would never dream of killing an otter themselves. What seems barbaric today was considered perfectly fair and acceptable fifty years ago. The otter was seen as an agile sporting adversary. Sometimes they were given a head start, and were not always killed when caught. Some wily and well-known individuals were respected and left alone. Where in the bad old days the otters were killed every time, and perhaps sometimes speared to death, more modern twentieth-century hunts avoided automatic killing and had long abandoned the tactic of spearing otters that was glorified in some Victorian art. Edwin Landseer's famous depiction of the climax of an otter hunt now serves as a chilling reminder of days that are over. In *The Otter Hunt* (1844) Landseer evoked all the gaudy action and glamour of a battle scene. The struggling otter is raised on a skewer by a sweating hunter, and a half-eaten salmon lies wasted on the riverbank, justifying the sadistic actions of the hunt. At the time, the scene portrayed a murdering otter caught and punished; to modern eyes the image now looks like some kind of strange ritual killing.

In the twentieth century, sometimes a hunted otter would be trapped in the water by a line of people forming a barrier with poles. This was called a 'stickle', and if caught like this the otter was less likely to escape. Williamson observed this when he followed hunts, and describes the method when Tarka is hunted. The otter could be deflected, or stopped from escaping seaward, and herded back upriver so that it could be hunted again another day, thus doubling its value in terms of fun and excitement. Stickle poles were trophies in themselves, and usually marked with notches for the number of kills the owner had witnessed.

Other more grisly trophies were taken from the dead otter's body, and these can still be seen in the country pubs where hunts gathered. The otter's tail was frequently kept, or its head was mounted, with a suitably ferocious snarl curled by the taxidermist into its varnished lips; its entire body might be stuffed as a memento, and even the trophy of the dog otter's penis-bone, the *baculum*, would be kept and sometimes used for a tiepin.

Northumbrian William Turnbull kept written recollections of his own otter-hunting exploits, and published them in 1896. In his *Recollections of an Otter Hunter* he lovingly describes his favourite hound 'Bugle', in passionate but questionable verses of poetry. The otters that Bugle found and killed were described in less favourable terms. One was described as being 33 lbs (15 kilos) in weight. This ferocious 'Fishmonger', 'Fish slicer' or 'Freebooter' was hefty enough to knock Turnbull into the water while it was bolting from the hounds. Turnbull describes the wiles and tricks of the quarry, and corroborates that an otter would not always be killed, but 'bagged', shown off, weighed and then released for a further day's hunting.

In his chapter 'Otters in Unusual Places' he reveals that they 'are quite familiar with the most various of retreats, eyots, old pollards, mill wheels and outhouses. There is scarcely a spot which will not harbour them, from a town sewer, to a garden, or even a shed.' He quotes a vivid report from the local paper, the *Newcastle Weekly Chronicle*, which described the watery movement of the otter which 'gangs like a ghaist, that swims in water as though it moved in oil'.

Ecstatic crowds gathered to watch the glorious hounds chasing the otter: 'The hunters, cheering with deafening cheers, made the woods and valleys ring and echo.' The baying of the hounds was known as 'music', and as they followed a fresh otter trail on the water surface, the dogs were said to 'speak' to the scent. In accounts of hunts the otter was sometimes accorded a heroic status, and as Turnbull describes the pitch of excitement during the hunt, rhetoric runs away with him and rhapsodic epithets such as 'sable diver' and 'king of the flood' pour over the pages, with the hounds 'like Tartars, tearing for his blood'.

The otter's new predator is cars. With the slow but steady increase in otter numbers, and the increase in traffic and roads, more and more otters are found dead at the roadside. As they disperse to breed, search for new hunting grounds, or move to avoid rivers in spate, their way is blocked by our noisy, petrol-scented strips of asphalt. That burnt, chemical stink must surely smell of danger, but it does not deter them if there is no alternative to the quick dart across. Millions of years of evolution do not include a memory of cars. Otters are not equipped for this new feature of human existence.

Otters move rapidly, and will always choose the quickest route across a road. Unfortunately, the time of day that otters often move around to feed coincides with dawn and dusk. This matches human activity and, during the winter, the morning and evening rush hour take place in the dark. Drivers cannot see the otter in the road until it is too late, and there are many more casualties at this time of year. Many of the otters that die on the roads each year inadvertently give their bodies to science, and I've decided to meet the group of scientists from Cardiff who collect them for forensic examination. The Otter Project is not generally known about by the public. This dedicated group run an otter research programme which depends on the financial support of various outside agencies. The Environment Agency pays for any otter that is found dead in England and Wales to be frozen and sent to Cardiff. The last time I saw James he was going to a rendezvous with the Cardiff people, with six otters frozen and bagged individually in his car boot. 'I don't know why we have to bother with expensive packaging,' he says, 'they've been in the freezer, they'll keep each other cold on the journey. It's only two hours. All we have to do is throw them on the train, and at the other end they can swing them out by the tail, slap them onto the slab and have a look. It could be done quite cheaply.' Although otters take two days to thaw out, the Environment Agency spends money on protecting railway passengers' nostrils from the aromas of thawing carrion, generously providing airtight polystyrene otter-packaging for the purposes of transportation.

It's late November when they can fit me in at Cardiff. I've only just got back from Cumbria, and the prospect of more driving is not attractive. I book an off-peak ticket and relax on the three-hour train journey from Devon. Gazing out at the unfolding serenity of the Severn Estuary and its surrounding fields of lapwings, geese and starlings, I attempt to blot out thoughts of what is about to be my first ever autopsy.

In Cardiff I find my way from the station through chilly streets and parks full of fallen leaves to the University Biosciences building. I'm met by Sarah, a young research student in a white coat. We go up several flights of stairs and through a maze of corridors into the nerve centre of the otter research project, and Sarah introduces me to Dr Liz Chadwick, who runs the programme. First there is a fortifying cup of tea and then we get to work. I'm bustled into an operating theatre and Liz looks for a lab coat for me that doesn't have too much blood on it. With a sinking feeling I realise I'm going to be doing more than just observing.

Dr Liz tells me they currently have a backlog of a hundred dead

otters stored. That is a lot of freezer space. Liz needs to go on mater-
nity leave soon and must pass on as much experience as possible to
Sarah, who will be taking over when she goes. This means that they
need to cut open as many otters as possible each day. 'How many?' I
venture. 'As many as we can stand. So it's good to have some help.'

I have mentally prepared myself for this moment. I know I am
about to see a dead otter and, aware that a first post-mortem experience might be outside my comfort zone, I have done my homework.
I looked at the papers Dr Liz and her team sent. I read about biliary
parasites and scent communication analysis and much more besides.
Along with scrutinising the photographic illustrations and evidence
that Liz provided with her reports and papers, I have trained myself
not to look away from gory images. I've kept my eyes wide open at the
moments on television vet programmes and episodes of *Silent Witness*
where the camera closed in on the operating table, on fresh organs
and muscle tissue. I've looked unflinching at photos of otter bile
glands and livers, and at a pair of unborn otter foetuses lying curled
like two perfect rubies on a glass slide. I have, I thought, prepared
myself thoroughly and professionally.

But I'm unprepared for the stench. ('You might want to take off
your jumper,' Liz advises. 'The smell sort of lingers.') Trying not to
gag, I do up the metal poppers on my lab coat one by one by one.
Fixed neatly and firmly up to the neck, I notice the old blood on the
front of my white coat, and the fresh blood on the floor. In my throat
the aroma of blood and fur and sinew begins to sting. It scorches the
back of my tongue and will stay with me for days.

Ahead of me, on a rectangle of zinc, in a lumpy and malodorous bin liner, is our otter. Together with Liz, I lift it. It's as heavy and awkward as any dead body, and as we cut the bag open the newly thawed otter pours unceremoniously onto the cutting slab. Its fur smells of old leather and grass, and undeniably composting organs.

'You're lucky, we got you a nice fresh one,' Liz says pleasantly. 'This morning's lot were not so good.'

Grateful that she chose me this 'nice' one, I concentrate on my clipboard.

'They can come in any state, from almost intact, like this, to mush. That can mean really stinky. When they're like that we can't find out much from them.' Liz adds: 'With the really old ones, it's sometimes hard to tell how long they'd been lying by the road before they were found.'

Earlier, we had walked past the room where the 'mushy' otters were cut up. Some scent molecules drifted out into the corridor; I tried not to inhale, but it made no difference.

Carefully we uncurl the otter and smooth it onto its back. Its front paws drop submissively onto its chest. Liz hands me a record sheet and a pencil, and as she measures the length, head-body and tail, I take the notes she dictates, tick boxes and add the figures. Writing down the facts helps me as much as it aids Liz, so I focus on the activity rather than on how I'm feeling. It is a dog otter, and fully grown; it is 125 cm long and weighs nearly 10 kilos. With the otter like this, its teeth show trickles of blood. His whiskers, or vibrissae, are snipped and handed to me, and as I seal them into a polythene sachet I notice how bloody they are. I try to breathe through my mouth only.

Liz checks the otter's limbs for breakages, gently moving the legs and rotating the joints like a doctor looking for fractures in an injured person. Her tender movements remind me of the doctor who examined my daughter when she broke her wrist. As I watched my daughter's arm being moved back and forth, I knew that the injury would heal, but my heart was in my mouth nevertheless. No injury will heal here, and inside my body, in my bones, I'm haunted.

I note everything that Liz says on the record sheet, pleased that my handwriting is small enough to fit in all the detail, and that I'm fast enough to take it all down. Having studied otter anatomy, I can spell most of the technical terms correctly. This small detail is like a life raft in a sea where I could easily go under.

Crushed skull. Broken jaw. Broken hind left leg. Teeth intact, with slight wear, and claws intact, except for one broken claw, front right, third digit. I look at the paws, the webbed toes, the limp leathery pads with their idiosyncratic wrinkles. Otters, I notice, have really big feet. The deeply wrinkled feet look worn, like the feet of an old person who has walked for a long way. The toe pads are surprisingly bulbous, as if they have been inflated. The tail is impressively thick, this rudder which could not steer its owner out of the stream of rush-hour traffic.

And finally, it's the fur, the thick, chestnut-brown fur, so rich and glossy that it opens an ache in my throat. The injuries, we conclude, are consistent with being bowled over by a vehicle moving at speed; a warm, soft animal kicked by a bumper, losing contact with the ground and meeting it again lifeless.

The vast proportion of the otters that come in have been hit by a car, Liz tells me. It's rare to find an otter that has died of something else. Sometimes they get a pregnant one, or a lactating female, and a cub or two that have died with her at the roadside. If a lactating female is found, and the cubs not, it has to be assumed that they have perished as well. During post mortem, uterine evidence will tell us how many cubs there were, but their whereabouts and fate usually remain a mystery.

Interestingly, many of the otter fatalities that are picked up off the roads are young, often under three years old. You can tell an otter's age from its teeth, Liz points out. In older animals the teeth become increasingly worn, but also, she tells me, the teeth have seasonal growth rings, like trees. These details help with forensics, perhaps even telling us what the otter's diet has been. She shows me where some otters have strange wear around the enamel at the base of the teeth, and I'm reminded of my own teeth.

Next Liz shows me how to make little foil packets. We are going to weigh and store the otter's organs. I watch out of the corner of my eye as Liz begins to swab the neck of our dog otter, and, using her scalpel, finds a space in the fur at the throat. She opens up his front from the neck to the tail, peeling back the pelt as if he's being undressed. 'It's great to get one in such good condition,' Liz says, revealing a fine layer of pale fat beneath the skin and the thick, dark muscle beneath that.

Inside the otter is a factory of organs undulating with reds: scarlet, deep plum, maroon. These are the scent glands, these the testes, kidney, liver, bile ducts, lungs, adrenal glands, muscle tissue. Everything must be scrutinised under the blade, anything at all must be noted – bite marks, parasites, subtleties in the organs – and everything recorded, numbered, weighed, packaged and retained in the freezer. This otter, otter number 1816, has minute kidney stones. Apart from that there seems to be nothing unusual, but even so, the body parts that are kept are invaluable for research. The skull of the otter is removed, cleaned and stored for research. We can't keep it; because of the otter's special protected status you need a licence to keep any part of an otter. Years down the line, the accumulation of this evidence might show us something we overlooked, evidence of some subtle change in the otter's environment that we missed at first.

Nothing else takes me by surprise, in the end, not the length of the gut as it ribbons out of the abdominal cavity, nor its contents (a small last supper of frog and tiny fish), nor the wobbling bladder nor the striped trachea, nor even the way we drop the used pelt into the incineration bin at the end. Nothing except the liver, which is so slippery it moves unexpectedly as I carry it to the weighing scales. And the heart. The heart, a shining dark maroon, cupped in my hands. It is like a miniature planet, the size and shape of a small clementine. I place it on the zinc surface of the scales to weigh. I don't want to let it go, this small, perfected heart, but in the end I have to.

East

The mind, full of curiosity and analysis, disassembles a landscape then reassembles the pieces – the nod of a flower, the colour of the night sky, the murmur of an animal – trying to fathom its geography. At the same time the mind is trying to find its place within the land, to discover a way to dispel its own sense of estrangement.

Barry Lopez, *Arctic Dreams*

It's mid December, and cocooned in the smart upholstery and polished chrome of a hire car, I'm heading north-east. I've been in contact with an otter man who seems to have seen more wild otters than anyone else I've found, and I want to know how he does it. I stop off to do some writing in Cumbria, shop for Kendal mint cake and collect the forgotten acorns I left at my mother's house near Cockermouth. They'd been waiting here for weeks, while I networked from my computer in Devon, hoping the weather would abate. Held for now inside their warty cups, these acorns will, I hope, produce some green shoots in the spring. They'll come with me to Northumberland, wobbling like talismans on the dashboard, out of the sleet and into the wind.

Driving east from Carlisle, I'm leaving my comfort zone. The landscape opens out into wind-blown hills, bare sycamore trees and scattered buildings of austere, dark stone. At the dramatic high sill of Hadrian's Wall, I wonder how the Romans experienced the threatening skyline, bristling with bad weather and blue-faced Picts. I have some well-heated armour in the form of this nimble little car and these days the locals are mostly friendly; it looks like the only thing I have to fear is the weather. With the mountains of Cumbria behind, the snowy backs of the Cheviot Hills rear up on the northern horizon. The map shows a mesh of enticing rivers flowing down from the Cheviots, through the Tyne's valleys, through rivulets, burns and streams and out to the North Sea. These are the icy veins and arteries that fan out to the eastern edges of northern England, and I hope they will contain more evidence of otters than the inundated west.

When I look at the body of Britain on the map, I sometimes imagine it as the shape of a bird in profile, and the region I'm in now looks like the jutting breast, a wishbone set just beneath the jugular throb of Edinburgh and Glasgow. Much of the edge I'm heading for is coastal plain, exploitable land that has been farmed, mined and altered for generations, and not one bit of it is untouched. The wildlife has either adapted or left. But thousands of migrant birds arrive here from Scandinavia and Siberia every autumn and winter, settling in the slivers of wild spaces amongst the graceful estuaries and dunes of the Northumbrian coast.

The 2010 national otter survey found that otters are doing better in Northumberland than anywhere else in the north of England. They retreated to the hills during the worst period of industrial pollution, and almost melted away. But it's a story with a heartening ending. A gargantuan clean-up effort was followed by a period of slow regeneration in the landscape, and in many places the water system is purer than it has been for a long time. Evidence of wild otters pushing back into their old territory can be found even in the built-up urban areas, from Sunderland and Newcastle to Blyth and Berwick-upon-Tweed. I want to see how a landscape that has been so intensely abused by human activity can recover and provide some sense of harmony with the wild. The presence of the otter shows that good health and vitality have returned to the ecosystem here. The restoration work has been costly, but it is a story of redemption if ever there was one.

I follow the A69 through a stark landscape of grey sky, green, undulating upland, bare sycamore trees and dry stone walls. Below is the winding, sparkling, muscular artery of a river. This must be the river Tyne, flowing along the deep valley it has cut for itself through the rock. Remnants of Hadrian's Wall jut out of the natural contours, and the high shoulders of the Cheviot Hills hold up a pewter sky heavy with cloud. This cold, beautiful landscape wakens my senses, and as the trunk road crosses a bend in the river I pull over and breathe in the surroundings.

Here two great rivers, the North and the South Tyne, crush together and begin their blended journey east towards the urban areas of the coast. I cross a sturdy bridge supported by nine stone arches straddling the broad rush of the waters. The water level is swollen, and the colour of dark brown bottle-glass. It froths over a weir and an inundated island. Salmon and sea trout can pass here, and many must be still pushing their way upstream. These fish are closely monitored on the Tyne as they indicate how healthy the water is; some may have just completed their lengthy migration from the sea, passing this weir to complete their life cycle, finally reaching the spawning grounds in the tributaries where they were born. The fish provide a superb food supply for otters, who have fattened themselves on their rich flesh for millennia. It goes without saying that this spot is also popular with anglers, but as salmon stocks are declining, most of the fish that are caught are returned to the waters.

I drive on and find myself in a market square dominated by a medieval abbey. This is Hexham. The sound of the name announces some kind of elemental defiance. These rugged borderlands endured many violent episodes over the years, and the feuds, uprisings and battles seem to have soaked into the landscape. During the Roman occupation the natural geological features were used as a stronghold to hold back the Picts, and in the seventh century, long after the Roman retreat, Wilfrid, Bishop of York, built a Benedictine abbey here, using many of the ruins of the old Roman buildings. When the Vikings invaded, Halfdene the Dane burnt down the whole lot, and only a singed Saxon crypt remained. The remnants of history still show in the abbey's walls, in a patchwork of Roman, Saxon and medieval stonework. The last Abbot of Hexham was hanged in the archway of the abbey in 1537 because of his resistance to the dissolution of the monasteries. Tragedy, devotion and heroism are everywhere in the surrounding land. Remnants of Roman roads crisscross with pilgrim routes, and pathways offer passages to the river now crossed by tarmac; the shallows have had fords and crossing places for nomadic humans for millennia and for wild animals for thousands of years before that.

In the town, all the voices I hear have rich, rounded vowels that resonate with the past and its Nordic leanings. I talk to a ruddy-faced dog-walker and question him about the river and its otters. He is well informed; otters have been hit on the road here, often coming out of the water when it rises in winter. And it seems the battles are not over; flooding has haunted the town's history since records began. The many tributaries that flow into the Tyne come through Hexham, and the magnificent bridge has been rebuilt at least three times. In 2005, exceptionally high rainfall caused so many broken trees and debris to pile up near the town that an important pipe was severed and drinking-water supplies damaged. At one point 10,000 people were without mains water. My companion furrows his brow as he remembers the devastation in the villages upstream. Since then state-of-the-art multi-million-pound flood-alleviation schemes have been used to protect Hexham. Solar-powered CCTV spies on flooding hotspots where the many burns flow into the town, and floods are predicted and prevented where possible. Otters and fish suffer as well, but special ramps and passes have been built for both to make sure that, in theory, we can live together safely. My friend gives me directions to see the stone facing and cobbles that were used to coat the leats and culverts between the houses where the rivers have been diverted. I'll have a look, but it occurs to me that this engineering activity may have been one extra pressure on the struggling otter population as it re-established itself.

In a few days I'll be meeting Kevin O'Hara, the otter man from the Northumberland Wildlife Trust. Kevin has an almost superhuman ability to sneak up on otters and has emailed me enviable photos which he has taken from high in the branches of trees along his local riverbanks. Also, because of his work as Wetlands Conservation Officer, Kevin gets to see all the angles. He visits sensitive wetland areas and consults with developers; he has expertise in every sort of wetland wildlife issue, from urban watercourses – where he advises on new developments and how to remove disastrous ones – to restoring wildlife havens and monitoring species. The Environment Agency has had its work cut out with Northumberland and its floods. And Kevin has had his work cut out convincing those in authority that wetland restoration is far better for flood alleviation than maintaining drained land, and that rewetting areas should not be seen as flooding, but as rehydrating. Where human needs are considered, short-term aims win, and always take precedence over the wildlife. 'The natural environment has become marginalised,' Kevin observes. 'It's viewed as a luxury that we only concern ourselves with in times of prosperity. But the environment and wetlands in particular underpin that prosperity and our well-being, regardless of what we think is important.'

With someone like Kevin pressing for sensible environmental decisions, I can see why there have been so many wildlife successes in Northumberland, and I can't wait to meet him. Wherever there is new development, it seems like Kevin has been there too, making sure all wildlife, from otters to bugs, water voles, shrimps and fish, is considered and allowed for. And, Kevin tells me on the phone, many new wildlife havens have been created. Marsh harriers have returned to breed here after 150 years of absence, and the most northerly avocets in Britain have also found a breeding site. When every inch of the land is spoken for, space must be made for nature, and the idea of Living Landscapes is prevalent here too. Living corridors will be created so wildlife can move freely and adapt more easily to human activity.

Rivers provide ready-made corridors for otters, as long as they are not concreted, straightened or culverted. But the wildness of water, its unpredictable, unstoppable force, means that as human populations expand, modernise and encroach, we have to look very closely at how we live. Solving environmental problems requires creativity and sympathy for this unpredictable, elemental mesh. Research shows that certain kinds of tree are more effective than anything else at slowing down rain run-off. Trees can massively reduce downstream flooding, as woodland promotes flood infiltration into the soil. It seems that the wild's requirements can concur with ours: the answer to flooding is more trees, and also the rewatering of ancient wetlands, which provide flood storage.

At Prestwick Carr, close to the airport on the edge of Newcastle, is one of the country's last remaining raised mires, and it provides just such a solution. Raised mires are mounded layers of peat that rise above the surrounding contours, sometimes several metres above the level of groundwater. They are useful because they receive rainwater and store it, as well as being a precious and unique habitat for wildlife. Eight hundred acres of this rare habitat are about to be restored, re-creating a huge sponge, soaking up rainwater and capturing carbon. The bog was formed after the last glacial period when depressions left behind a mosaic of swamp, meres, ponds, bog and fenland. After

many years, dead plant material built up to produce this rich formation. It was drained until it was bone dry in the nineteenth century,
and many rare species disappeared; if the water is retained here once
more, it will provide flood relief for nearby towns and restore a rich
wild habitat for thousands of species.

I drive past some more miraculous upland geology towards the city.
This is the windswept volcanic outcrop called the Whin Sill, with its
eroded, rippled walls and plunging valleys gouged out by ice and
water. I know from Kevin that the Tyne is doing well for otters. Newcastle is now celebrated for its green credentials and its recovery from
the industrial years when the river was ecologically dead. Even in the
built-up parts, where it might seem wildlife could never live, otters
have been attracted by the cleaner water and habitat restoration.
Where the river flows through the city centre salmon and sea trout are
back, and otters have been detected from the suburbs to the estuary.
Otter cubs have been reported in the heart of the town; spraint has
been found beside the footbridge over the river, and on the quayside
close to the shopping centre. Otters have been seen in the river near
the pubs and shops, and even caught on CCTV stealing into suburban gardens and investigating goldfish ponds.

Skirting north of Newcastle, I travel east to where the land stretches out, widens and rolls gently to the edge, the rich inter-tidal zone
and the sea. Beyond the road, there are startling white dunes, miles of
them, up and down the coast. And eroding these is the wind that
constantly flays the aptly named 'bent grass' which grows along the
edges of this coastline. The flashing wildness of the North Sea looks

out east to the huge, dark horizon where, generations before, the Viking ships appeared. Now the invaders are of the winged variety. I can hear them flying in by the dozen, on the cold breast of winter. Multitudes of birds are coming, and the sky is echoing with their calls: whooper swans, geese, redwings, waxwings, snow buntings. The sky is full of wind and wings. Many thousands of these birds land at night, and if people could see their spectacular arrival they would stop in their tracks and marvel.

At the village of Beadnell flocks of gulls flap beside my car, dipping into the shelving outcrops of sandstone at the sea's edge. I can see the flash of their eyes and wing tips; they are up and over the wind, over the splashing waves and over the car, disappearing and reappearing as distant flecks of white against the black cloud. The tiny Northumberland cottage where I'm staying, lent me by my friend Alex, is only a few yards from the sea. Stepping out of the car, I can smell whipped-up spray, salt and weed, and miles and miles of wind-blown sand. Battered by the wind, looking west, I can see the high hills of the Cheviots with a covering of white snow. The noise of the sea is huge and overwhelming. I light the electric fire as it gets dark and spend a night under a rattling roof with the roar of the North Sea in my ears. It feels as if the waves are right at the door. As I dissolve into sleep I dream I'm in a whirling boat and wake on the edge of my mattress as if I've been blown there. I fall out of bed, expecting to see a flood tide as high as the window. There is nothing of the sort, but the grim daylight hasn't brought better weather. The wind is just as strong, the sea is still at its roaring best and the old sycamores, the only trees which appear able to withstand the winter battering around here, have let go of all their leaves and their branches scratch wildly against granite walls and dark sky.

The next day, I find the burn on the edge of the village. This little otter way has a strange, ancient-sounding name. 'Long Nanny' is thought to be an old Celtic phrase for stream, and it twists out of farmland and into the sea, mingling the scents of sheep pasture and old watercress with the salt and weed washing over the high dunes of the bay. The wide curve of sand and mountainous dunes look as if they have been bitten out of the coast, and the stream's mouth is hard to locate at first. When I reach it, it seems to have shifted its position with the movements of the seawater. The bay bends and smooths all that I can see, and reputedly hides the bones of rusting machinery beneath its flattened surface. I've read that anything that lies buried here will resurface again at some point. An otter was found dead in the village, not far from the beach. It must have come out of the Long Nanny, perhaps on its way to the next burn flowing out to sea at Seahouses, a mile further north. When food is scarce in these little streams, otters use them as routes to the seashore, only coming out of their secretive paths if they're blocked by debris or made impassable due to flooding.

The wind is relentless. It scorches my ears, and the only protection is the fish houses on the stone quay. These sturdy buildings used to be lime kilns but now fish boxes and fish smell fill them and seep into my nostrils, while I shelter from the fray. A gap in the sea wall reveals

tremendous brown waves and clouds of spume. Gulls shriek overhead, surfing the salty updrafts and dodging the waves. I leg it across
the sand with my hood flattened over one side of my face. The sky
begins to throw down hail, and with the squall comes a lightning flash
that lights up the slate-grey underbellies of monstrous clouds. Some
say that the otter that was found dead had been shot, others say it was
hit by a car. One person confided that some people around here still
view otters with hostility, and thought it was a deliberate killing.
Whichever it was, it may have been human-induced, and even one
loss like this can affect a fragile otter population.

I stay long enough to ascertain that Long Nanny is a tiny burn,
tidal and muddy, but big enough to support some sticklebacks. I
search for signs of otter but find none. Inland, beyond the mountainous tons of sand, the burn cuts through a muddy set of fields and
passes what looks like an ancient standing stone. Further on there are
gnarled, windblown hawthorns and an old leat running to a ruined
building with walls crumbling named Tughall Mill.

While I am here, I visit the Northumbrian poet Katrina Porteous,
who lives in a cottage by the sea. She writes about the old wisdom and
seafaring traditions that have been dying out in her part of the world.
The herring fishermen who fished the coastline near her house are a
threatened species, and she saw their gradual loss as terrible. She
writes about the men and their boats in her book *The Blue Lonnen*.
The book is an elegy for the wooden boats and the people who were
the keepers of both the old Northumbrian language and a deep
knowledge of the seabed:

Their marks,
Their grid of bearings, like the stars —
Not just a map, but a mesh of stories —
Lit up where and what we are.

The small harbour shelters a few brave boats. One is a Northumbrian coble. I've read about the boats in Katrina's book and it seems they're special. They were crafted from wood and built for sailing, but with the coming of the engine it was easier to rely on petrol and satellite navigation, and they ceased to be built. The East Anglian artist James Dodds has evoked the boats' sympathetic magic in a series of paintings which illustrate *The Blue Lonnen*. In some of the pictures, the boats seem to be suspended in mid-air. One is painted from a fish's-eye view and captures the boat's smooth underbelly. In Dodds' painting we cannot tell whether the harmonious background of resonant blues and greens is sea or sky, only that the keels of the boats float in some transcendent place, in the realms of the past and of memory.

Herrings are no longer caught here, as the seas that were once rich in these fish are only slowly recovering their stocks. When fishing was at its height, otters may have felt the added pressure of competition, and now that it has gone, they, like the fish, are returning. But the cobles, and many of the people who built them, are no more.

Beyond the harbour and the beach are rugged stone outcrops, a good feeding ground for otters in winter, when crabs, eels and shellfish are exposed by the rising tide. Otters and men once lived shoulder to shoulder on this coast, both with a profound knowledge of its contours, currents and dangers. Upstream, inland, fields of grass green the landscape. Once the turf was ripped up for industry, and littered with furnaces and coal pits, but all that has gone now.

Katrina tells me about a map of the seabed, hand-drawn from memory by generations of local fishermen. It has the names of the rocks, reefs and troughs of the hidden undersea. So familiar were these people with the contours of the fishes' world that they knew it all in their heads and could recite it by heart. They were as at home at sea as on the edge of the land here at Beadnell. Traces of this wisdom are still written in the ground in a pathway of broken mussel shells that Katrina calls the 'blue lonnen' in her poems about the past. The well-worn road, lined with cracked shells, led from the fishing huts to the sea. Here, the hand-built fishing craft were taken down to the water to find herring. Otters have paths like these lonnens and will use the same ones through the generations. When the otters declined, their paths grew over. But many are reappearing and now once again you can see their sinuous ways through the grass by the streams up and down this stretch of coast.

The air is chill and vibrant. Today I'll drive to meet otter man Kevin. I spread the map on the bonnet of the car. So many rivers! Many of them, all across Northumberland, from the Wear in the south to the Tyne, Blyth, Wansbeck, Coquet, Aln and Tweed in the north, seem to flow roughly (some with spectacularly elaborate meanderings) from west to east, many pouring in parallel into the North Sea. Kevin didn't seem surprised by my questions on the phone about otters in the area, and when I arrive at the rendezvous at Druridge Bay, there he is, ready in his wellies and combat trousers. He grabs my hand and shakes it so heartily that my whole body feels warm.

It's a good start. Kevin tells me humorously that otters make themselves unpopular locally with the cohorts of birders by eating the rare geese that blow in from their migration on the East Atlantic Flyway, the migration route used by 90 million birds annually. He explains that it may have been easier for otters to spread here not just because of the food supply, but also because of the underlying geography. He draws a grid in my notebook, to demonstrate. Otters can pass easily over the coastal plains, he shows me, moving his pen between the main rivers and through their many tributaries. The waterways have formed a navigable network, and look like a geometric grid of paths from the higher ground in the west to the plains in the east, the finer tributaries providing the north–south paths and major rivers the east–west corridors. Where there is less connectivity, the otters may use the sea to travel between the mouths of tiny streams and at the same time use the rich coastal resources to boost their feeding. Otters may have been able to spread more easily through this type of landscape than those further south, or inland, where many areas are blocked by hills, mountains and roads, or simply the impossible engineering in the waterways of some cities. The national otter survey bears this out, as otters are doing better here than anywhere outside the south-western stronghold where I live.

Geography may be one reason otters have been so successful, but even more important has been the grand scale of the clean-up following the decline of the opencast coal mining, shipbuilding and steel industries. Industry devastated many parts of the landscape during and after the Industrial Revolution. No part of the landscape was untouched, and the land will never return to its original state. But there has been a huge amount of long-term restoration. And the geology may have been part of what made a difference to the otter's recovery; here in Northumberland many of the rivers are young, shallow, fast-flowing and dynamic, cascading off the granite of the Cheviot Hills over rock that is very close to the surface. These rivers renew themselves remarkably quickly.

Some of this local rock, Kevin explains, the dolerite, also known as whinstone, was molten lava which flowed over the surrounding Carboniferous rocks about 300 million years ago. The waterfalls, pools and rippling shallows of the burns which flow from these upland rocks contribute huge amounts of unpolluted water to the weightier rivers like the Coquet and the Tweed and attract more salmon and sea trout than in other areas.

In a wildlife-watching hide that looks out over an old mining-subsidence pool at Druridge Bay I learn that many pools were left behind after the mine closures. The old opencast mines, which followed the deep mines and dotted the landscape, had a finite lifespan. When they were abandoned by the coal company, the deal was that they should be restored for wildlife and landscape value. The coal company would fund this work, and in conjunction with the Wildlife

Trust's expertise, many new wetlands like this one at Druridge Bay have been created. The Wildlife Trust has managed these new wetlands. Kevin tells me about the reed beds, and how, as time went by, restoration practices improved. At first the edges of the pools were planted geometrically, and their edges were too straight. Now the pools have more natural-looking curves and undulations. They are easier on the eye and provide better habitat, but also stand up to the elements more effectively. To begin with, reeds were selected from the south, but they were not up to the challenge of the cold northern winters and they perished. Later, reeds from Yorkshire were tried; this hardier variety survived, and as ecological succession took over, habitat developed, and a home for a whole range of aquatic life was established, from invertebrates to waterbirds, raptors, fish and finally otters. One of the pools has been designated a Site of Special Scientific Interest just for its shrimps, Kevin tells me.

The Great Whin Sill with its rugged escarpment not only inspired the Romans, it is also a cradle of wildlife. The Romans made use of the sill to augment Hadrian's Wall, but otters and other creatures had been exploiting this underlying geology and its rivers and streams for millions of years before that. Research has shown that the geology of an area can significantly influence the otter population, and in coastal areas, if the rock is impervious and provides freshwater pools, there are far more otters. One study, carried out by Paul Yoxon of the International Otter Survival Fund on the Isle of Skye, found significantly more otters in southern Skye, where the rock is Torridonian sandstone. The number of otters increases on coasts with this type of impervious geology because it provides the vital freshwater they need to feed, drink and clean their fur. Many of the pools here in Northumberland are entirely man-made, but have been a boost for otters nevertheless. The network of clean water flowing from the Northumbrian uplands means otters can travel vast distances, with ranges from the rich, hidden wetlands of the blanket bog around the granite summits of the hills all the way down to the sea, where they feed, drink and, according to Kevin, pester the seabird colonies.

Kevin and I walk on the dunes and talk about otter behaviour. Kevin has been watching otters and working to conserve them for many years, and observes that the otter behaviour handbook does not exist. They defy all the rules, he says, and he suggests that they are far more versatile than I had thought. Some of the things he has observed contradict what others have seen. He describes often watching otters in broad daylight. He has observed them encountering one another, and thinks they could be less territorial than we generally assume. He describes watching otters going wild with joy on finding one another, as if they are less solitary than we might think. If their ranges overlap, he points out, they may well be related. Even more surprising, he has seen dog otters left 'in charge' of older families of cubs while the mother leaves for long periods to feed, almost as if the male were babysitting. He describes seeing otters hunting on land, stalking voles forced out of their burrows by building works, and bringing back mouthfuls of the voles live and wriggling for the waiting cubs to practise their catching skills. The dunes here house many rabbits, and in winter when fish are less plentiful the otters spend a good deal of time finding and eating them. It sounds quite different from the shy otters I have tracked and observed in the south. Kevin once photographed a dog otter that was so curious about the man in camouflage hiding in the old tree that it swam as close to the riverbank as it could in order to peer back. Once, an otter came out of the water, trotted up to Kevin and nipped his boot. Another, on the dunes, found his sock lain out to dry, decided it was a spare one and stole it.

Back in my cottage I meditate that perhaps the country contains many otter tribes, some of which have their own behaviour and culture, according to the landscape, food supply and geography of where they live. An otter in Devon, it seems, has different and perhaps more secretive ways than those in Northumberland.

A mile or two up the coast from Beadnell at the fishing port of Seahouses, there are plenty of fish and chip shops where I can refuel after my day out with Kevin. On my way there, the road crosses several tiny streams. In many places these isolated streams that flow into the sea are known as denes. If it is true that otters have been using these streams to get down to the coast, soon, as night falls, they could appear. I have to decide on a spot where I can watch them, and if Kevin is right, I could see one anywhere, capering over the golf course, rummaging through somebody's washing or hunting rabbits on the dunes.

Eating my chips at the harbour, I spot some new-style fishing boats, very different from the traditional old wooden cobles, with up-to-date satellite navigation and many-horsepower engines. These are large enough and comfortable enough to take tourists on sightseeing trips to the seabird colonies on the Farne Islands. A few miles out to sea, this dark archipelago of dolerite has cliffs with jagged names like 'Fang' and 'Knivestone'. Recently, an otter was found there. When I say 'found', I mean it was discovered to be living there, but typically of an otter, it was not actually seen in the flesh and fur. I wonder how long it stayed? Was it was trapped there alone and lived out the rest of its life, well fed but isolated?

The blustery Farnes house thousands of seabirds, and might seem inhospitable to anything or anyone else, but not perhaps to an otter. One thousand years ago St Cuthbert stayed there as a hermit. During his stay, probably because he had no one else to talk to, he took pity on the wildlife. The stories tell of Cuthbert's closeness to animals; he had befriended otters while on the mainland, and medieval images depict him standing on the shore with otters playing at his feet. He also protected the local population of eider ducks, who were known ever after as Cuddy's ducks. I can imagine him, one cold spring, adopting a nest of eider eggs and nursing them, between prayer and contemplation, until the hatching stage, when they may have imprinted on him as their parent and followed him everywhere. At the time, some may have thought that his miraculous relationship with animals was God's work, but idle gossip suggests that he may have been covetous of the ducks because of the lovely warm duvets of down they produce. Whatever the reason, the outcome was the first ever official protection for birds, as Cuthbert blurred the boundary between the wild and the civilised.

Why would an otter brave the perilous distance to the Farnes?

Typically, he was never seen, but his prints were identified in the mud at the water's edge of one of the outer Farnes. The otter must have endured cold, unpredictable waves and near exhaustion to get where he did, and he would have been very hungry when he got there. The 80,000-strong avian population must have offered a fabulous smorgasbord, and some recompense for the journey. In all he swam three miles from the coast, and this staggering feat of strength and resilience astounded locals and experts alike – otters had never been recorded there before. If ever proof was needed of the astounding wiliness and strength of the wild otter, this is it.

Not surprisingly, no one will take me to the Farnes in this weather, and I settle for standing like the French Lieutenant's Woman, gazing out to sea. A fulmar rises out of the cliffs, its wings outstretched like a little albatross; it climbs spirals in the wind, eyeing me warily. I walk onto the short grass at the edge of a golf course near Seahouses, wind in my hair, and look out. The distance to the islands is not inviting. But if the otter population is growing in this area, young adult male otters would need to find new territory. This might be the reason for the otter on the Farnes, or it could also be that the otter was caught in the Force 9 winter gales that occurred just before its tracks were found; confused in the waves, it may have headed for the dark shapes of the islands, thinking they were land. With its nose at water level, it might be possible even for an otter to lose its way in choppy conditions. I imagine that otter – if it is still there – lying on its back in some dry, sheltered cranny, its belly swollen with a feast of crabs, shellfish, herring and seabirds.

I scour the coast for any sign of otters, from Beadnell to Lindis-farne, but all along the ragged, icy edge of the land my eye is dis-tracted upward. The sky is laced with winged migrants scudding in from the north and east. The voices of geese, the flicker of snow-buntings, the magnificent honk of whooper swans; every piece of wind seems to echo with bird sound. Wind has sculpted the stunted sycamores, hawthorns and whinny bushes, shaped the dunes, tum-bled clouds, blown hats and whisked hairdos into the sky, but every-where it is the birds that astound me. I turn into a shaft of low sun-light and a raven flies right in front of my face. It lands on a wall and eyes me, the purple sheen of its back and wings dazzling in the pol-ished, wind-whipped, salty light. A volley of starlings shoots by, as if they've been fired in unison from a single bow. A field shuffles with curlews and dances with lapwings and, on the sand, turnstones, plovers and oystercatchers rush about in front of the spume.

Just outside the town of Seahouses there is another burn that trick-les down to the sea. After a long inspection I finally find something. An otter has been making use of the golf course to cross from the burn to the sea. There is enough spraint in the grass by a bridge over the tiny coast road to suggest that a family of otters might be in resi-dence. They could be roosting close-by in one of the bramble patches or reed beds, just out of golfball range. I can see where they exited the water, followed one another up through the grass, over the road and on towards the putting green and the dunes on the other side. Early-bird golfers, their hats tied onto their heads, are making the most of the sunshine and have no inkling of the otters that have been cavorting over their neatly mown lawns. The dunes are a wild, dynamic environment and must provide good rabbit hunting during the winter months; I find a fresh corpse lying unfinished in a crook of the sand, and it looks like something has been gnawing on it recently.

While most of the landscape from the hills to the sea has been exploited and spoken for, the dunes have never been much practical use and form a kind of last wilderness. A pristine strip between the land and the sea, they are constantly moving. I remember as a child the prickly excitement of dunes; the tufts of strange grass and the mountainous shiftiness were dynamic and full of potential for adventure. The dunes were ancient, and occasionally let go magical clues to the past. Fossils of huge Scots Pines would occasionally work to the surface, hinting at the prehistoric Caledonian forest which remained buried for thousands of years; as the dunes eroded, the spontaneous appearance of these lost relics was like a message from some hidden underworld. Wherever dunes arise, they support rare species; lyme grass, bloody cranesbill, sea lavender and creeping willow can grow here, and these attract rare insects, plant hoppers, grass miners and shore flies. Many species of bird feed here, as well as small mammals.

I find some more spraint close to the dunes, on the smoothed carpet of a putting green. Following a trail through marram grass I discover the gathered couch an otter has made not far from the burn, just where it parts the dunes: a small bed of turned grass and some dried bracken where an otter has curled and slept under the sky.

When I remember that the forecast is for clear skies tonight, I'm taken with the idea of staying out. Before long I've packed my winter sleep-ing-out equipment, determined to spend at least some of the night in the wild architecture of this sand-scape. In my five-season sleeping bag, in full view of the sea and the burn rushing over the sand, I may hear or see the otters at dusk or dawn. There are signs that they pass through here almost every night, and I want to lie in the relatively sheltered quiet of this place and hear and smell some of what they would. These sand-scapes are made out of longing and loss; they are like a living brain, remembering and forgetting what the wind and the sea have made of them. The dunes are gradually spreading inland, and could swallow roads and farmland that lie close-by. Further south, in County Durham, the mounds of ugly spoil from the opencast mines near the coast were like shadows of these pale ones. In the past the unsightly mountains were gathered up and taken out to sea, where they were dumped. But the ocean currents brought everything back, with the silt and sediment rearranged and rewoven in dark, swirling gradations, like beautiful strata of the sea's choosing.

Soon the moon will be casting its light over the water and out to the horizon. I gather some driftwood and settle into a spot where there is a hollow in the sand. I make sure I can still see out, and sit listening to the incessant rumble of the rollers, waiting for the light to change. Around sunset the wind drops a little. I light myself a little driftwood fire. The stars come out, and as the sky deepens I can see that they are not simply brilliant white sparkles, but many colours: pink, blue, white, red and yellow.

The contours and folds of the dunes around me take on the texture of dark velvet. A huge dome of stars revolves over my head. I put on gloves and a hat and get fully clothed into my bivvy bag, filling it with a scattering of sand. No matter. It is dry and all will be well. Although I want to look outward, my inner compass draws me to the glow of the fire. The moon rises behind me. My ears pick up more and more noises: the burn doodling a thin, musical trail of light; the relentless roar of the waves; the breath of marram grass. Once, then twice, I hear a thin whistle, carried above the wind, but in spite of myself I feel sleep drawing down my eyelids. I curl close to the embers of the fire to warm my face, and fall asleep in wafting aromas of woodsmoke and seaweed.

Some time later I'm shaken by a shock wave of air beside my head. In the moonlight I catch sight of a bright shape; something has landed beside me. The white-rimmed circle of a face where there should be no face looms into focus. The moon has come down to meet me, but it has claws, ferocious yellow eyes, a hooked beak and black, dilated pupils. An owl has come to my camping spot. Beneath the blur of flecked plumage, clenched in the dark talons, is something small and dead. Before it became prey, the poor mouse or vole was foraging in the treasure trove of my biscuit crumbs, having the last supper of its little life.

The owl cocks its head at me, and I daren't breathe. I'm quite sure it can see far more of me than I can of it. With my face poking out of my bag, and my wide eyes, perhaps I look like a large owlet. Its metallic, shadowed eyes pierce into mine and spook me. I've never seen such an owl before, and the image of this haunting face scorches itself into the back of my eyes. As it takes off, moth-like, revealing its broad, pale under-wings and dark wing tips, it floats bodily into me. I can't sleep after that, and return to the car to sit through the small hours, protected from the wind, waiting for the light to return.

On the east coast, a clear dawn often arrives with particular acuity. At other times, the sun can rise over the sea casting horizon-wide reflections through the haar, the low sea fog that can linger all day around here. In winter, even when the North Sea is slate-grey and choppy, there is a moment where the light lifts itself from the water and for a time the wind and salt cease to hurt the eyes as the light grows by subtle grey increments, hefting the sea-damp and darkness away.

When the sun comes, rising slowly from the direction of the sea, I creep down to the water's edge and wait on the sand for otters, but either they saw me first or the unexpected gift of the owl was all I was granted.

With fresh spraint in the marram grass, and paw marks deeply imprinted all round, I can tell that an otter has passed close by during the night. The narrow passage of its body has left a trail through the grass, with moisture still settling in the indentations and in the crumbling sills at the end of the prints. But I'm cold. My hands are white, my fingertips hurt and my feet are losing their feeling. The otter will by now be curled in its pelt beneath the gorse or amongst the reeds, and it's time for me to re-enter the realm of the human. The raw longing in my belly for warmth and a steaming cup of tea sends me homeward.

Travelling down the eastern edge of England, my journey south takes me across the Tyne Bridge. The city that grew up around the Tyne was once synonymous with shipbuilding and steel. The urban skyline contains this memory in its contours of old buildings, renovations and new developments. The last coal cargo left the Tyne in 1998, and now service industries dominate. Today Newcastle and Tyneside are gaining recognition for urban regeneration; they won a prize in 2010 for the country's greenest and most sustainable city. For me, the miraculous return of wild otters to the river is capped by the magnificent city skyline around it: wildness is blending itself in. Kittiwakes nest amongst the towers of the renovated Baltic Flour Mill, and the innovative design and sympathetic adaptations of The Sage, an international music venue, are reflected in the river. The Sage's spectacular contemporary silver-glass curves can also be seen from the inside in a magnificent concourse that looks like the inner belly of a whale. The Gateshead Millennium Bridge, with its 'blinking eye', built for pedestrians and cyclists, is the world's first tilting bridge. This graceful, almost delicate-looking structure takes less than five minutes to rotate its semi-circular form and tilt 40 degrees upwards, thus allowing small ships to pass underneath. Hence the name 'blinking eye'. It even cleans up its own litter. But it isn't just in recent years that the environment has been cared for; the ancient green tree-filled common land at the Town Moor has never been built upon, and this area is bigger than London's Hampstead Heath and Hyde Park put together.

The river Ouseburn, a tributary of the Tyne, runs through the parklands of Jesmond Dene in a series of waterfalls, pools and wildlife

havens which are perfect habitat for all sorts of wildlife. One primary
school, above the old shipbuilding yards at Walker, asked its children
to conduct a survey of all the animals the children had seen within a
one-mile radius of the school. The results revealed rich and diverse
animal life returning in and around the river; more than sixty species
were seen, from large emerald moths, to trout, salmon, crabs, eels,
jellyfish and seals. Many new housing developments in the city now
include Living Landscapes; they have sustainable, energy-saving technology, car-pool schemes, charging points for electric vehicles, and
large areas given over to woodland and meadow, all criss-crossed with
cycle ways and footpaths.

Following a lead, I drive deep into suburbia and park outside a
modest semi in a street of semis, to meet a lady who has been talking
about something extraordinary in her garden. Jeanne told me on the
phone that she blamed the cat first, and then the heron (which she
never saw), and then somebody said it could be owls, but when she
sat up for a few nights and watched, she saw a visitor that she never
expected.

Jeanne makes me a cup of tea and we sit inside, viewing her small
fishpond through the patio window. After an hour and a half, not
long after dusk, in the sodium yellow of the street lights, a flash of
caramel fur catches our eye. One minute there was nothing, and
then, with a shuffling, curious gait and a highly whiskered muzzle,
comes sniffing an obviously hungry mustelid. An otter! Jeanne has
put out a little fresh trout for encouragement, which the otter
munches with gusto. As we watch its poor table manners through the

glass, I can see it's a fully grown otter, and it is doing its best, licking its lips after the first morsel, to lift the otter-proof chicken fencing that has been fixed down over the pond. It's looking for more. It pads around the pond, pausing, sniffing, scratching, searching for a way in. Any remaining fish must be quivering at the sight of the huge predator pacing the pond, trying to prize its fangs underneath, digging with all its might at the fencing. After a few minutes of savage digging at the wire it finally accepts failure and gives up. Through the double glazing, I catch the silvery front of its dry fur, and then the unmistakable *huff* as it decides there is no more to eat. It lifts its tail and leaves a spraint, then squeezes back beneath a dark gap in the fence and is away to hunt in the river, or perhaps in somebody else's pond. The sighting of so large and mysterious an animal in a place as incongruous as a small, paved suburban garden shows that otters, like urban foxes, are truly adapting.

If you leave Tyneside by road, and drive south through Gateshead, you can't miss the strange sarcophagus figure of the Angel of the North. Her wings are outstretched as if guarding the country of the North East. The huge, rust-coloured wings are made of steel, and as I drive, I notice her ribs, curved like the inside of a ship or the half-made fuselage of an aircraft.

The wind blows flocks of birds over the city. They flow in from the sea, surfing the aerial currents, dodging Gormley's tall sculpture. This haunting, bird-like statue seems to fuse together the industrial, the brutal and the angelic. Flying around the solid structure, the birds seem fluid, sailing high over the city, following ancient weather patterns, aerial highways that have been here for millennia. On the radio I hear this modern angel's foundations being discussed; the roots may be concrete, but they stretch back, further, through the layers of history and memory; through battles, farming, steel and coal, factories, to as far back as the time of St Cuthbert and beyond.

I'm struck by the alchemy of the human and the geographical here. The wild weather and light, the struggles, the culture, the contours. Those gone before would all surely take pleasure in the news that the mesh of all these connections has not been lost. The otter, the dingy skipper butterfly and the many rare species of bumblebee that illuminated the patterns in the manuscripts of the past are returning to newly replanted wildflower meadows on old industrial sites, on stream sides and riverbanks, and in parks across the city.

Driving away from this cityscape where the wind batters relentlessly from the North Sea, it is the image of the angel and the architectural jewel on the banks of the Tyne, The Sage, which linger in my mind. It's true that the regeneration they represent connects old wisdom with the green of new growth. And threading through it, with cleaner water from the springs in the high hills to the streams, rills and soakaways in the towns, through the freshwater leats and regenerating subsidence pools, through the many tributaries, into the rivers and down to the sea, is the irrepressible, dynamic return of the wild otter.

Holt

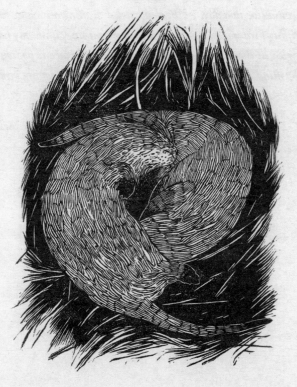

*I could live two days in the den, curled, leaning on mouse fur,
sniffing bird bones, blinking, licking, breathing musk, my hair
tangled in the roots of grasses. Down is a good place to go, where
the mind is single. Down is out, out of your ever-loving mind
and back to your careless senses.*

Annie Dillard, *Living Like Weasels*

While some of the rivers in the North East are winning the accolade 'pristine', the privilege of such clean water does not extend to the area I want to explore next. I'm heading towards where I was born, and where my love of otters first began, but here in the South East, with intense pressure on watercourses and an expanding human population, rejuvenation is distinctly uneven. There is thought to be a population of about fifty otters in Northumberland, but further south the latest national otter survey was hard pressed to find anything like this number, drawing blanks in many parts, especially in south Essex and London. None at all were found in Kent nor in my native county, East Sussex. According to the latest report, the few wild otters that remained at the edges of these areas have continued to suffer from the stress of human presence, and any road casualties or incidences of toxic leaks or poisoning had catastrophic consequences on a fragile otter population, leaving surviving otters isolated, unable to find breeding partners and left to die out.

I want to look at some of these most troubled rivers and find out if there is hope for the otter here. The existence of otters in Newcastle is so encouraging that I can't believe the story in the South will be completely gloomy.

The 2010 national otter survey found signs of otters on the Thames as far east as Windsor, but no further, even though there are several areas that should be habitable. The London Wetland Centre at Barnes provides a giant banqueting table of over a hundred acres of rich watery landscape for birdlife and potentially many animals that need marshes, water and fish. But otters need connectivity. They can't fly over one of the biggest conurbations in Europe to feed and breed. Otters can only spread if they have safe, clean, otter-friendly corridors through which to travel. Through London, there are slivers of wetland and marsh, but these are isolated islands in a sea of urban development. Many of these reserves are cared for by local communities, small trusts, and sometimes larger organisations like the RSPB and the Wildlife Trusts. An umbrella structure, Water for Wildlife, supports, encourages, advises and coordinates the work of many of these bodies, but as no otters have been seen in any of the London rivers for a very long time, a clear set of requirements needs to be met before otters will consider moving back in.

Work *is* taking place to clean up the water and provide sheltered hiding places on the riverbanks, but many city waterways wouldn't be able to support otters at present. In the Thames region, massive investment would be needed, particularly in the sewerage system and in sewage works, if stringent new European water directives are to be met and water quality improved. The potential is there; think of the other species that were extinct in England. The common crane, which stands taller than many of our children and has a wingspan larger than an eagle, had not been seen here for 400 years. During the 1990s, in an effort to attract the rare bittern to the East Anglian Fens, an area of carrot fields was transformed back into natural fenland at the RSPB Lakenheath reserve, and some cranes found their way there, without human help. Once the invitation had been issued, that intimate connection of creature and habitat was rekindled naturally.

The Environment Agency recently found 119 species of fish surviving in the spoiled grandeur of the Thames Estuary. These were mainly coarse fish, but salmon and trout can also be found in the Thames, not only in the headwaters coming from the chalk and limestone aquifers, but swimming upstream from the sea. The good news is that the river Thames won an international prize in 2011 for improvements in management and restoration. This is a river that has been to hell and back, from being declared biologically dead last century, to being clean enough in recent years to support salmon and sea trout.

If we imagine rivers like the Thames as arteries fed by many tributaries, and think about what we put down all the drains that flow through our city and its parks and our own gardens into the rivers, we begin to see what is to be done. Little by little, we can clean them up. Take the river Lea, or Lee as it is sometimes spelled. This river is loved and abused in equal measure, and has attracted attention because where it flows through London, it arrives at one of the biggest new building sites in Europe. Government regulations now stipulate that any development close to a river must consider its ecological status, mitigate for wildlife and plan for its regeneration. The lamentable state of the river Lea didn't escape the Olympic developers' notice, and its restoration has been put at the heart of the new Olympic Park. The vast scale of the Olympic development has attracted massive investment in a coherent plan for the natural environment, and the river nestles in the middle of it all, repercussions rippling up and down its length.

In many parts of the river, from its source to its mouth where it pours into the Thames, the Lea is being restored. The Lee Valley Nature Reserve now follows the river in its upper course, and the Lea River Park takes the river through its more urban sections in London. Otter holts are being built in secret locations, an indication of the optimistic hopes for this river.

In the 1990s a number of captive otters were released in the upper Lea to aid the ailing population's recovery, and although some disagreed with the policy of reintroduction, it is thought that a surviving population of otters in this rural area upriver now averages about six. What will happen to these otters? What their chances are, and how they can be helped to spread, depend upon what we do to restore their habitat.

At its source in the Marsh Farm estate in north Luton, the river Lea is at its furthest point from the new developments. At Leagrave Common, water pours out from the bedrock of an underground chalk aquifer hidden somewhere beneath the Chilterns. At this point,

things don't look great for the Lea. The source is protected by a concrete tunnel and a metal grate which prevent a build-up of rubbish
from blocking the flow, but untold toxins have already seeped into the
water. Warning signs declare the watercress that grows here unfit for
human consumption.

The water begins its patchy journey, slipping between urban high-
rise car parks and roads, making the sound a young stream should,
whispering between the reeds that push through a boggy patch of
marsh. The water soon comes into contact with beer cans, trolleys
and plastic bags, but pushes on, travelling through fly-tipped rubbish,
leaving Luton and meandering to more rural parts. The river flows
through a pastoral section, amongst grazed fields in Bedfordshire and
Hertfordshire, and later it forms the boundary between Middlesex
and Essex. At first its waters are clear, and in spite of the rough start,
a variety of plant and insect life seems to thrive around it. But the
reedy meadows, trees and grassy banks are quickly replaced by a busy
human environment, which continues and builds in intensity all the
way into London, where the dense population means that its water is
estimated to be about 50 per cent processed sewage.

Another hindrance to the river is its base level – that is, the rate at
which it flows out of the aquifer – which is at the very limit of the
human population it can support; the river is being sucked dry.
Despite the hosepipe bans in recent years, many people continue to
drain the water reserves by digging their own boreholes. If the whole
region is 'water stressed', it follows that if any more new houses are
to be built, the aim should be to significantly reduce the amount of

water each person uses. Putting aside the issue of water conservation for the human population, is there any hope at all for wetland wildlife when habitats like this have been so exhausted?

Yes – in spite of everything, nature reserves and repaired wetland habitats have sprung up along the length of the river Lea. From Walthamstow Marshes to the Cornmill Dragonfly sanctuary, from the Waterworks Nature Reserve to Tottenham Marshes and Rye Meads, all the way to the mouth of the Lea near Bow Creek Ecology Park, the river passes by and through areas that are being cared for and mended. Even though large stretches of the Lea have been canalised, flooded with sewage and poisoned for generations, and in spite of the years of East London's infamous 'stink industries', the Lea is set to recover. When this recovery will be fully achieved is another matter, but many authorities have at least put their minds to starting it.

From Limehouse to Bow Creek there is a chain of navigable channels and locks, but against the industrial skyline, reed-bed restoration and tree planting are in full swing. On and around the island that is home to the new Olympic Stadium, rapid change is taking place, in spite of the toxic combination of contaminants that have been found in the water and surrounding soil. Heavy metals such as lead, arsenic and chromium have had to be cleaned away, as well as ammonia, fuel oil, tar, bitumen and chlorinated hydrocarbons. Other problems have been landfill, agricultural run-off, urban drainage, sewage works, soap factories, gasworks and many of the other industries that have taken their ruinous toll. We know the effects: in the waterways fish have been failing for many years, plant and birdlife have suffered, insects, otters and water-voles have disappeared. The whole natural ecosystem has been in a state of collapse.

While the construction of the Olympic site takes place, it doesn't look nice, and it's hard not to feel depressed. But this river is in the middle of a huge transition. We have come out of a trance and, having opened our eyes, we find we care about it after all. Many local people mind that there are very few dragonflies or flowers. We need clean water as much as wild nature needs it, and we want it. With good legislation and investment, it's possible to clear up some of the mess. The new plans will bring huge improvements. Where the river Lea flows past Stratford, the wasteland and the river clogged with stagnant duckweed and lumps of sewage look bad on the aerial photographs of the new Olympic Stadium site. Moreover, the new European Water Framework Directive has changed the measures by which we judge water quality. In a wonderful ecological ripple effect, many of the brooks and tributaries that carry sewage into the Lea will have to be cleaned up. Where before we only tested water quality for some chemicals – we concentrated on ammonia levels, 'organic solids' and dissolved oxygen – European law now requires we test for about thirty things, including the river's whole biology. We have to look for living organisms, for diatoms (phosphate-indicating algae), fish and invertebrates. The key indicators of a river's health must be what live in it, and this can only be good for wild otters.

At Stratford, a few miles away from where the Lea spills into the Thames at Leamouth in London's East End, the river splits into several channels. One of the world's most advanced, sustainable, urban drainage systems is being built here to support the development around the stadium. Alongside the eight kilometres of waterways, a new wet woodland will be created, with 2,000 trees and over a quarter of a million wetland plants. Where the river Lea passes the gigantic space-age structure of the stadium, and further upstream, otter holts have been built. Frog ponds, grass-snake egg-laying nests, rare bumblebee habitats, bat boxes, loggeries, moth strips and water-vole and kingfisher habitats have been created. Newts have been removed for their safety by hand, and will be reintroduced when the time comes. Hand-mixed and hand-sown native wildflower meadows will be planted, and it just happens that the predominant colour when the Olympic meadow bursts into bloom will be – you guessed it – gold.

Beside the stadium, over the razed rubble of the past, human homes have also been constructed; they overlook the river, and a new street has been named 'Otter Close'. Perhaps one day, as well as the magnificent contours of the stadium, the silhouettes of real live otters will be seen from the windows of these new des res apartments.

The critics say that while the Lea has received attention and investment, rivers like the Wandle in south-west London have been neglected. The Wandle is a good example of a river we pass every day without noticing it, because in London so many tributaries of the Thames are hidden from view. They flow submissively between concrete embankments or inside tunnels deep underground. These rivers, covered as sewers long ago, live secret subterranean lives and where they burst forth they are often seen as nothing more than drains.

The river Wandle, which gives its name to Wandsworth, was for a few hundred years treated like a sewer, but its health has come full circle. Like the Lea, during its long relationship with the people who settled around it, it has been poisoned and abused, but in spite of intense urban pressure this once lovely chalk stream now supports wildlife again. The river may have attracted less attention than the Lea, but it has an equally interesting history.

In prehistoric times the river flowed in a pristine state from its source high in the Surrey weald, and continued nine miles to its mouth in the Thames. Rainwater falling on the wilderness of the North Downs percolated down through the chalk, and reappeared as a beautiful, clear jewel of a stream, verdant and throbbing with wildlife. As the human population expanded, the fast-flowing stream's energy was noticed and harnessed. The Domesday Book recorded thirteen mills along the Wandle, and it was later made famous by Izaak Walton, the angling writer who wrote meditations about catching its trout. Before urban sprawl engulfed it, it was an iconic trout stream, and Lord Nelson fished here. As late as the 1820s, John Ruskin, who visited it as a child, described untainted parts of the river as it progressed 'under the low red roofs of Croydon, and by the cress-set rivulets in which the sand danced and minnows darted above the springs of Wandle'.

There were once ninety mills situated along other parts of the riverbank; there were calico works and flour mills, mills for snuff grinding, for bleaching, tanning, copper milling and beer brewing. At one time the river ran bright with the reds and blues that were used in the tanneries. By the 1960s, the Wandle was officially declared a sewer, and much of the river was paved over and forced into concrete embankments. But in the last ten years green oases have sprung up, from Waddon Ponds in East Croydon, to where the river resurfaces at Beddington Mill. Where there were once working corn mills and tobacco was brought to be ground into snuff, the minnows are back and a population of brown trout, nurtured by local schoolchildren, has begun to thrive. Thames 21, an environmental charity, has been mobilising people to clean up waterside grot spots where rubbish has built up. All over the city, volunteers – mobilised by the Wandle Trust and other small organisations – are coming out of offices and schools to spend some time in the sunshine, clad in neoprene, to wade about in the alien world of mud and silt, removing plastic bags, traffic cones, tyres, rusty trolleys, old scooters, discarded clothes and old iron.

Fish in the Classroom, an initiative that puts small hatcheries in schools, delivers fish eggs to local primary schools and children are asked to care for the young fish that hatch. Through the glass of the new aquariums, children can see into the underwater world of sticklebacks and trout and witness their development. Later, when the small fish are ready to be released, the children make a trip to the local waterway to liberate the young shoal. One teacher involved in the project noticed that her class became sensitive to the responses of the fish to their environment. When their sticklebacks were becoming stressed, the children felt protective and responded with a respectful hush in the classroom to soothe the fish.

It's January and very cold when I explore the grey-green band of parks that stretches along the Wandle. The river crosses the four London boroughs of Croydon, Sutton, Merton and Wandsworth, and like the Lea forms a wildlife corridor linking each to the other. Areas such as Elms Pond and Wilderness Island provide habitat for tufted ducks, kingfishers, woodpeckers, nuthatches and treecreepers. At Poulter Park and Watermeads, the water is clear and greened with weed and last year's vegetation; reed mace, reed sweet grass, nettles and creeping thistles support a vast array of insect life. At Deen City Farm the river flows twenty feet wide in places, and further downriver shoals of dace can be seen, along with grey wagtails and many other species of birdlife. On the narrow stretches of the river in Wandle Park and the Wandle Meadow Nature Park moorhens hide amongst hemlock and water dropwort. In spring, the walls of the river are thick with pendulous sedge, male ferns and hart's tongue ferns, and further towards the Thames marsh yellowcress will flower and garden angelica will provide insect fodder. Eels and pike have been seen here, and coots and moorhens nest amongst watermint and brooklime. Where the Wandle meets the Thames, smelt, dace and flounder spawn, and terraces have been built to draw in an array of waterbirds such as the charismatic great crested grebe.

Although there are still modifications to be made to the bankside and its concrete relics, the Wandle may one day take its place again as London's finest chalk stream. Those young people who released the fish that they reared in their classroom might one day witness a water vole plopping into the water as they pass by, and their children might even find the tracks of an otter by the river.

I continue my odyssey by travelling from London to the far south-eastern corner of Britain, to where I grew up in East Sussex. An exciting phone call came from my father, to say he had seen an otter swimming up the river Ouse in my home town of Lewes. This muddy river was where my childhood search for otters began, and there had not been any sign of them here for decades. I listened excitedly to my father's story. Out shopping early one morning, he was crossing the bridge by Harvey's brewery and was distracted by a wake moving in the water. He peered down at the milky brown surface of the Ouse to see a dark creature swimming past, nose aloft, clutching a fish in its jaws. An otter in Lewes? After all this time! 'Phone the Wildlife Trust and tell them,' I ordered, and hastily bought my train ticket.

I made my plans before I found out the stark truth – that although there is a sprinkling of otters on the borders of the county, there are no otters at all in East Sussex. One grey February day, I phoned Fran Southgate, the otter woman at the Sussex Wildlife Trust, hoping against hope that she would have some better news. No, my father must have been mistaken, she said. At the Wildlife Trust they have been scouring the county for signs of otters for many years, and although there have been one or two hints on the borders, and habitat restoration is under way, otters haven't made it back into East Sussex yet. I looked out at the cloudy sky feeling depressed. What my dad had seen was much more likely to have been *a mink*. As in many parts of Britain, this species has been rampaging its way through the water-vole population and reducing it to tiny pockets of resistance in one or two places. When later I met Fran in Lewes we talked about this some more. If the otters were established here, she explained, they would help to control the mink, acting like guardian angels (mostly) for the poor water voles.

Fran has been working on the Sussex Wetland Landscapes Project, compiling a survey and map of all the wetland habitats in the county, and the results have been sobering. The rate of wetland loss has been monumental but has gone largely unnoticed. The luxuriant fens, rare chalk streams and ancient floodplain woodlands have largely disappeared from the Sussex landscape, along with the many species that inhabited them. The absence of the otter, at the top of the food chain, is a key indicator of the losses. As the landscape was drained for agriculture, more roads built and human developments expanded, the wildlife vanished. Google Earth and modern electronic mapping technology have provided exciting tools with which to look at our landscape, but what they do not show is what was there before.

In Lewes I plan to visit the chalk streams I played in as a child. The Cockshut and the Winterbourne ran past the bottom of my garden and I spent hours there, fishing for sticklebacks, collecting tadpoles, staring at the alien faces of dragonfly nymphs through the glass of a jam jar and discovering the horrors of leeches on bare legs. When I get there, I realise that these days most of the streams in the area are victims of 'watercourse modification'. They have been stopped up with tidal flaps, impeded by concrete flow-obstructors, straightened with embankments, or blocked by road development; their once-crystal-clear chalk-stream water is full of duckweed and litter. A short walk along the Cockshut, where I used to roll up my trousers and paddle about with a net, shows water that is barely moving, and on the banks people have tipped old plastic chairs, discarded roofing felt, and left piles of glass bottles, mattresses and old bedsteads. Where the water is flowing more freely, it is canalised by high, bare embankments, or even worse, in the town, the Winterbourne is encased in solid concrete. These concrete tunnels were exotic when I was a child, a great place to play; there were no leeches, and no weed to slip on. Playing here was like entering a science-fiction world where our make-believe was enhanced with clinical edges and corners; the ground was predictable and the water contained. We didn't give a thought to what had been lost.

At the edge of Lewes I walk along the river Ouse and peer into its silty depths. The river is tidal here, and I can see ripples and rises on the surface which suggest that the river is packed with fish. When I peer into the confluence where the Winterbourne flows into the Ouse, the mixing of the clear chalk stream with the muddy tidal water is astonishing. It looks like a mingling of glass and clay, and in the marbled swirl hundreds of huge grey mullet and other fish species cluster around the outflow of clean water. But their passage is blocked. As with many of the chalk streams and tributaries, massive metal tidal flaps have been installed, and the fish cannot travel through. Engineering for flood alleviation takes precedence over the salmon and trout; it will take the citizens of Lewes a long time to forget what happened during the last flood, and few people would appreciate the removal of this important protection. When I talk to a neighbour who is interested in fishing, however, he remembers the day the Environment Agency removed thousands of wild sea trout that had come to spawn in the Winterbourne. Halfway upriver, they met with a tidal flap and became trapped. With nowhere to go they beached themselves and died, in huge slippery, silvery heaps.

Many years ago these chalk rivers, in their natural state, were allowed to meander and their gravelly beds were an ancient spawning ground for trout and salmon. They poured with pristine water and were nourished by the tidal salt marshes that reached all the way to the sea. From the top of the castle that stands at the highest point in the centre of Lewes, you would have been able to see a wide stretch of glittering marsh, and a mesh of natural channels reflecting the sky and the wings of a million waterbirds. Bounded on two sides by the smooth rise of the rolling chalk Downs, the wetland would have stretched to the horizon. The Winterbourne and the Cockshut ranged wild and free; wide and deep enough to be used by barges and sailing ships that came from abroad to trade with the town. The Cluniac priory at Lewes, which once looked out over the reed beds and marshes, was made from stone brought all the way from Normandy. This stone had been sailed six miles inland, up the wide channels that were once here, which are now blocked or filled in.

Before it was drained, the marshland was animated by reeds, pools and flocks of waterbirds: waders and ducks, lapwings and cranes, herons and swans. Their voices and movement filled the air. Today, human needs have meant these rivers are kept under the strictest control, and most of the birds have fled. The fluid cycles, the un-trustworthy change and flow that used to govern the marshy water system, were seen to be in conflict with the town's homes and businesses. I find myself longing for the marshes that were once here, and it seems like an impossible dream, but the rehydrating of the marshland has been discussed. It would be prohibitively expensive, and is generally viewed with fear and scepticism, but the idea of re-wilding parts of the area is creeping to the surface.

With their habitat blocked and their progress impeded by towns and the natural geographical barrier of the South Downs, otters have not yet been up to the challenge of recolonising. In the muddy old water meadows, at Lewes Brooks, beyond the roaring bypass, the RSPB has made a small reserve that boasts a sliver of habitat for lapwing, snipe and rare aquatic plants. The Sussex Wildlife Trust is also working hard to restore habitats like this, so that they connect up and spread, and future generations will be reminded that their rivers did not always flow between embankments and in straight lines.

On the old railway land on the outskirts of Lewes, a restored flood-storage area is being left to its own devices, and people walk beside the reed beds and water meadows, along the silt-green swirl of the otter-less Ouse, past the new Linklater Pavilion, which has been designed and built to accomodate floods. Walkers can meander like the river, to a mysterious spiral of pathways designed by the artist Chris Drury, named 'Heart of Reeds'. Local people walk their dogs or stand and view the living land-sculpture of reeds. Amidst the grandiose rolling scenery of chalk downland, in an organic seasonal rhythm, water is sluiced through the reeds and out into the water meadows, through a sensitive double-vortex which links the landscape with the human body; Heart of Reeds is inspired by the shape of the human heart.

When I heard that an otter was hit by a car on a road in Polegate, only eight miles from Lewes, my heart missed a beat. I thought that even though this otter was dead, its discovery might mean that otters had after all returned to this part of Sussex. I hoped that it proved the pessimistic reports wrong. Otters *were* here after all, and had been living undetected all this time. But this story is in fact worse than the story of the mink sighting. The road casualty was the first otter known about in the area for many years. The national otter survey, however, came to a very depressing conclusion. It reported that this was the only otter detected in East Sussex, and although otters could potentially spread eastwards from Hampshire and southwards from the Medway, the finding of the dead otter did not tell us anything new or exciting. It suggested the opposite, in fact. The otter in Polegate was possibly the only remaining survivor in the area. This lonely otter may well have spent its last months, weeks and days starving and searching for company. Repeating the pattern of extinction that happened country-wide in years gone by, this otter may have ended its short life looking for any small sign of another otter, never knowing that there was none.

Back home in Devon, I begin to look at my surroundings more closely. I feel reassured in the knowledge that there are otters nearby in the river Dart, but what about all the little creeks and tributaries that flow even closer to my house? I live on the edge of town and, in an uncharismatic and muddy little valley nearby, water springs from an aquifer deep in the rock. Where folds in the hill crease enough to spill out ground water, I look at where three trickles move together. They are delineated only by earth that is a little more spongy than its surroundings. Rushes and other marsh plants grow up like tufts of green hair. As the three streams gather into a proper watercourse, rainwater percolates down and joins in, and slowly, subtly, the valley begins. Where the rich, red clay contours dip sharply into the valley, and water cuts between the green, rounded hills of the South Hams, I would expect to meet nobody but a surprised badger or startled crow. But there seems to be an otter here. I've found some spraint by a stream just outside suburbia, less than a mile from my house.

There is a track, an eroded old green lane, and only the farmer uses it. The stream joins this track as it heads downhill toward the marsh and the river. The area hasn't been built on, and at night in the bottom of the valley, by the widening sliver of marsh, there is no light pollution. If you come here on a clear night and sit between the wavering reeds you can see the constellations glittering like the inside of a huge dome, as if the stars had never been outshone by the dazzle of our towns. None of the usual road noise penetrates this muffling landscape, and in the wetness of the valley floor a small and unnamed water course meanders freely in ribbons of reflected light.

The countryside is full of these secret rivers. From the hillside where it is born, this one seeps unnoticed alongside and beneath a road that takes traffic away to the coastal resorts of the Devon Riviera. From the road, the only sign of water is the rushes. Most people, travelling at forty or fifty miles per hour, wouldn't see it, especially since it disappears then reappears a little while later as a squashy mire. They wouldn't know that something is whispering to itself, and travelling down towards the deep water just beyond the hill. Even the river Dart is hidden from many roads; driving through Totnes, you barely catch glimpses of it, and where its waters widen through the hills towards the sea, you might only see its brightness as you approach from inaccessible villages like Dittisham or Stoke Gabriel.

Before it reaches the river, my stream is perhaps still only a foot's width as it wriggles between grazing fields blotched red with sticky clay. A few Red Devon cows plod down to drink the cool springwater. I am afraid of the curious young bullocks that are sometimes around; they canter cheerfully en masse, and I've once or twice been cornered and forced into the stream. If I see the cattle before they see me, I follow the old farm track which hides me from their sight and walk along it until the stream trickles alongside. Here I can enter the shallow water safely as it flows beside the track. Both it and I are hidden by a gauze of hazel trees. Grey squirrels skitter above my head and eye me from a safe vantage point in the tunnel of twigs. I had never considered that an otter would come here, but knowing what I know about otters now, I understand that they do not mind about scenery; the last thing they need is glamorous landscape – it attracts the wrong sort of attention. Otters need places where their presence can blend in; almost anywhere undisturbed, in fact, as long as the water is clean. This valley is perfect. I want to reassure myself that otters are resident here, so near to where I live, and I begin a more sustained period of tracking.

I go in the early mornings, while everything is frosted with dew. Spring creeps in by increments as the daylight hours become longer. These deep lanes respond to the increasing sunlight by pushing up green buds which day by day, in all weathers, increase their lushness. The celandine buds are the first to burst, and with a dazzle of yellow petals, they catch the light and shine like molten butter. Leaving the path I jump down into the water to cross the shallows and trespass into an overgrown meadow. Above the meadow, on the slopes of the valley, a great many trees have been planted, but lower down, brambles grow in thick clumps. In between, small stakes have been driven into the earth. Each one has a sign nailed to it directing those who are shooting to do so responsibly.

This explains the nervy hedges, crackling with the scurry of pheasants. The hen birds' plumage blends with the woven undergrowth. If you get too close they suddenly lose courage, panic and explode away from your footfall in an extravagant and heart-stopping flurry of clucking. The foxes are well fed around here. I often encounter one particular dog fox and around any corner might catch him sunning his fur, daydreaming about rabbits or vixens. Today I can smell his presence hanging amongst the yellow gorse flowers.

The foxes and the otters keep well out of the way of the shooting, but the otters do it best. There are no roads crossing their stream, or their river, only this quiet track, so they are safe for the moment, but even if they were within range I wouldn't see them. There is so much cover they don't need to flee noisily like the pheasants. They do exactly what otters do, slinking out at dusk, making themselves impossible to see, their fur the same colour and texture as the mud, banks and undergrowth.

I scramble through a gap in the hazel branches and land mid-welly in the water again. The gravelly stream meanders loosely this way and that, leaving unpredictable pools and shallows in its curves. It's a stream that doesn't have to be anything but itself, left utterly to its own devices, and it provides plenty of shelter for wildlife. There are rocks at regular intervals, each one a different character; this one bare and favoured by wild ducks, that one a chiselled plateau with beetling crags and jungles of moss. The stream edges rise and fall, making a series of dynamic environments; miniature cliffs that crumble and shift from year to year provide soft earth for water-vole holes, and tall grass and swamps intermingle with the tattered straggle of brambles and undergrowth. The irregular nature of all this makes it virtually impassable to humans, but an otter could move along here unseen. The tumbled rocks are just right for a series of sprainting stations. As I walk I search each one. At last, on the largest and most ostentatious rock island of all, I find some spraint. It might be a day old, as it is desiccated and cemented to the rock, but I can see the fading calligraphy of many other stains that suggest an otter's regular visits.

I survey the surroundings. All around the meadow there are hidden bramble patches and holt-like spots where an otter could easily be lying up for the day. On the hillside fifty metres away there are banks of roots and huge trees sheltering badger setts and fox tracks. Would it be a bitch or dog otter? I know that females bring their cubs up tributaries. These places are entirely different from the main river because they are so private. The water is shallow and slow flowing, the cover is thick. It would be an ideal haven for hiding young, with the spongy and absorbent marsh not far off; there seems to be little scope for interference, or the dangerous kind of flooding that would snatch and drown any new cubs.

The meadow comes to an end where the stream crosses the farm track. If I want to continue, I must cross where the water dips into the dark under a low bridge and heads into the tidal marsh. It's too low for me to swim through, but an otter might. I scramble down to the water on the other side of the bridge and find spraint just where it should be, on a rock where the otter must have left the water. There is also a rubbing place with a lot of stray hair, the sort a moulting feline would leave on a favoured cushion. But what cat would sleep in the wet, mossy armpit of a stream?

The wet pathway is only a few centimetres deep as it riffles away and disappears into the marsh. I make my way through into the overgrown reeds and my feet begin to sink through the surface. This is treacherous ground, unmapped swamp-territory. Strange drowned trees poke out of the mud, their bleached limbs catch the light like ivory bones. The reeds and sallow are soon taller than my head, and I am dwarfed, sinking lower and lower into the silt. It's easier to walk along the gravelly bed of a sister stream where the water has fanned into an unidentifiable number of trickles; the sound kaleidoscopes all around. This is moorhen and coot land. Their wide toe marks balance easily in the shifting wet. It's no good for heavy-footed intruders. This tangle has no path unless you are in the water at nose-level, and the marsh begins to unnerve me. An uncanny squealing makes the hairs on my neck prickle. It is not otter, and badger would not be out at this hour; listening to the scintillating array of voices, I press on into a world of waterbird language and alarm calls.

Where the reeds end the stream has collected into a vast, swirling pool behind an old dam wall and it is being sucked unnaturally down as if into a plughole, bursting through a submerged pipe and out of the marsh. At the point where the stream meets the banks of the wide, glittering river Dart, mud is everywhere. It glares and shines and dazzles. I crouch down to hear a thousand tiny mouths creaking with mud-sound. This is the view an otter might have as it came out of the marsh. At the edges, water and mud lose themselves together. Higher up, oak trees and rock are moored to the bank, and there are dry places between trunks to curl up. I place myself in the crook of a rock and a woody root, and my nostrils fur with earth-odours. If I were a hungry otter I would follow the rim of the crumbling wall and bank, blending myself with the contours, and later drop down a level onto the rocks to search for crayfish and eels as they emerge on the rising tide. I can smell otter in a tangle of fishy grass that lies pungently right beneath my feet. I move to the cover of an oak trunk and lean back. Right beside my seat the otter track comes out of the wide tidal stretch, just where the river curls in an elegant meander around the farmland and vineyards of the estate on the opposite bank. In the silt I can clearly see where the tracks begin on the tide line. The tracks move in a straight line up to the mouth of the tributary. For the first time, I notice there are two sets of tracks, and on closer inspection I decide it might be three. Some of the pad-marks are smaller. A mother and cubs!

The tracks move so closely together they mingle and cross one another, as if they were moving symbiotically. I can see the fusion of their mud-coloured forms moving together in my mind's eye, as they scurry back to the sheltering labyrinth of the marsh, the cubs following their mother's every move. The otters stopped to spraint here on the low wall, and disappeared into the reed beds in the valley bottom. Close to the tracks are those even smaller, scratchier marks that the light-fingered mink make. This is not the first time I have found mink and otter tracks so close together, and it intrigues me. It is as if the mink has been following in the track of the otter, maybe just out of sight, and perhaps by some clever trick disguising its own scent.

As the tide rises I wait. Time blurs, the water darkens and the day softens into something other than itself. Just as the light fades they come, on the threshold of dusk, out of the reeds, mud-coloured, so close to the ground in colour and shape they are barely distinguishable. My heart flips and my eyes struggle to focus. They do not sense me, but undulate over to the edge of the river, sleeking into the water as one. Whether they sense my presence or not, they are cat-footed and stealthy, with the bodily knowledge that danger could be anywhere.

I meld myself into the bank, trying not to breathe. A family of otters, so close to my home! I stretch up to peer as their heads resurface and the two cubs, three-quarters the size of their mother, move alongside her as shadows. Perhaps they know I'm here, the hint of my human scent drifting toward them.

A little while later, I decide to go in after them. I came prepared for this: out of my bag I pull a wetsuit, a hood and some water shoes. In a sheltered spot I shed my clothes and pull these on, stashing my dry things by the cliffs, and walk down to the water. I take the otter's way in, and slide in down the shallows in a slick of silt, head first into the water, into the brackish stench. Weed and mud waft into my face as the cold ripple of the incoming tide comes up to meet me. It lifts me into the cold, salty flow and seeps into my wetsuit, down my neck and all over my skin. For a moment fear freezes my limbs and air dislocates itself from my lungs.

The tide is drifting gently upriver. I relax a little as I feel the water carrying me. The suit makes me unnaturally buoyant, and I realise I can swim easily. Heading around the meander in front of me, nose at water level, I adjust to the temperature and move my legs. Otters learn to swim on the surface as cubs, and it takes them a while to learn the knack of tipping the nose slightly skyward, so the nostrils can still breathe. In this way, they make a ream of water in two lines that fan out behind them, like the wake of a subtle boat. Swimming like this, nose up on the current, takes hardly any effort at all, and below me, in this green-brown world, lie glutinous depths of eel-soft mud. I try not to touch its octopus-tentacle slime; mud and oil patterns slide over one another in dark slicks. I can't look at it, and when I feel things brush against me, whether it's the brown sides of fish, eels or evil-fingered drowned trees, I feel vertigo and cannot think. All the memory of the river is here, hidden, concealing time, stories, lives; old sunlight, life and earth rubbing up against one another in a rich, vibrant, dynamic muddle. I wonder if the otter can read all these layers, or if they are significantly different from place to place as it moves over the water-scape of the river in all its phases.

In patches the river is the colour of dark amber, moving constantly with reflections of geese, a cormorant drying its wings in a tree, flurries of starlings. I swim until I can swim no further, and haul out a short distance upriver, close to some woods on the edge of town. The otters are nowhere, but I can still smell them; the otter and cubs must enter and exit the water frequently on this stretch, to eat, spraint and aerate their fur.

I take myself by land back to my clothes and get dressed quickly. My whole body tingles with the freezing freshness of river. Walking on foot upstream in the gathering dusk, past the town, I aim for the towering weir with a large pool caught beyond it. It is near-dark now, and the town lights are filling the sky with orange. I plan to pitch a camouflaged tarpaulin shelter beside a thicket of shadows. I know a spot where I can use borrowed town-light to see the surface of the weir pool.

A few days ago I found several salmon corpses lying scattered around the weir, where they failed to make the leap on the journey to the spawning grounds. Chewed from the tail up to the neck, the corpses attracted gulls and were beginning to rot. I marvelled at one fish head, almost the size of a roast chicken. It would be far easier for an otter to wait for the powerful fish to die in this way and enjoy the free pickings, than hunt here. Many predators must hunt like this, going for the easy kill rather than wasting energy on a difficult pursuit. Otter spraint often confirms this, and around here, outside spawning times, it is usually full of the bones of small fish and eels that have been easy to catch. I know there is an otter holt hidden between some large rocks and sallow trees not far away. It is entirely out of reach from humans, being on the wrong side of the river, and although someone with a canoe could paddle across and get a closer look, nobody ever does.

My family appear. We have made a rendezvous and they have brought me hot tea, food and a sleeping bag. I sip the hot sweet tea and unpack my hide. I'm going to thaw, eat and wade across to otter country, staying where I will be in view of, but not too close to, the holt.

My hands and face regain their proper texture, and the light is gone. The skin of the water draws a glassy mist over itself, subtly tinted with sodium from the town's glow. I position myself in complete seclusion in some willow scrub to wait. This is generally a good time to settle into an otter spot and stay quiet. It's dark now, but in the borrowed light I may see them pass. In the morning there is a chance I could see the otters as they work the weir – although I've been here on many mornings, and evenings, all I've ever seen is a cryptic ripple as the otter disappeared downstream, or into the edge of the bank, and then nothing. Otters have a way of melding their colour to the water, even in daylight, and you can hardly make out the otter's body as it slips water-ward, catching only a ripple as it makes a crescent of light over the curve of its back.

With the damp river air flooding around me I sleep lightly. A female tawny owl cries '*kee wicc*' and is answered by her mate's soft reply, '*whoo ooo*'. Strange splashes and twitches punctuate the darkness, and I wake again and again. At dawn, my eyes are open at the very first peep of a stirring bird, and I untangle myself from the knot of my cocoon. I look out into a grey world. The river is ethereal in wild half-light. A coot calls through the mist. I can feel rain coming. Small, fine, sporadic drops at first, then gradually more and more. Covered in camouflage and totally waterproofed, I wait to see the otter. I can sense her nose, just above the water surface, sensing me. I scan the edges of the water with my binoculars. Where is she? I can almost smell her. Otters can remain hidden for as long as it takes, their nostrils just at the surface, drawing in the stray molecules of our scent. With mine hanging thickly in the air, my otter will not show herself. I really should be making tracks.

Caught between two worlds, I gaze at the glassy dawn river. I don't feel like going home yet, nor do I want to swim in this cold pool, even though the water is enticing. What is this longing to be near the river? At the very least we might share this one thing with the otter, this attraction to the comforting concealment it offers and the intoxicating lightness it gives. It is easy to forget that once we were at home underwater, and never truly lose our desire for it.

338

I stay by the river a little longer, but whatever I hoped to see, I could never have predicted what happened next. This spectacle is a very different sort of wildlife encounter than usual, and one so close that it does not require binoculars.

The grate of something moving along the riverside footpath on the opposite bank distracts me from the water. It is light now, although grim rain-light, and two bantering youths in hoodies are moving noisily along in my direction. What are they doing on the river path at this hour? It is so early in the morning I decide they must have been to some all-night celebration on the edge of town and are finally making their way home at dawn. They can't see me; I'm across the water and they wouldn't expect anyone to be there. I stay put, feeling outnumbered and self-conscious in my strange gear. I wish that they would go away. Just when I think they've gone, they burst back into view, loping straight toward my hideout, having removed every single item of their clothing, including underwear. I sit wincing at their brazen nakedness in these cold conditions. As their pale bodies hurtle into the pool, exactly in front of me, they whoop and splash, and I give up all hope of seeing my otter. The two swimmers haul themselves out and repeat the exercise, and I begin to wonder whether what I am witnessing is not entirely human, but some kind of strange shape-shifting. Too late I realise it is not. It dawns on me that when they catch sight of me crouched here in the bushes there could be some considerable misunderstanding.

When they spot me, they are as startled and put-out as any otter might be. They thought this water was theirs alone, and now their spell is broken. I can tell they are truly human, because instead of disappearing gracefully water-wards, they slither clumsily up the bank, thoroughly offended, grab their clothes and streak off in the direction of the town.

As for the otters, I don't see them again. The river must have called them away somewhere else, down over the weir perhaps, flowing with the live current in the water, to some new stream or marsh that I have yet to explore.

Acknowledgements

A vast array of people have helped and influenced the progress of my search for the wild otter. As there are many who know far more than I about this elusive animal, I want to apologise for any inaccuracies or mistakes that might have crept into the text. The strongest influence upon this book has been the otter itself. My learning about it will never be over, and the combination of its wild mystery, its total otherness, and the many voices of its landscapes continues to hold me in its spell.

I should go back to the beginning and thank my parents, Oliver and Shirley Darlington, for introducing me to *Tarka the Otter*, and for encouraging my interest in stories as well as in science and the natural world. I hope they know that the combination of poetry and science that they offered was the catalyst that drove me to follow the

341

wild otter. I also send my heartfelt thanks to Mark Cocker, who was present at the birth of my plan to write it all down in 2008. Along with Stephen Moss, Mark encouraged my quest from its first inkling and welcomed it into the fold with good humour and a keen critical eye. Mark agreed to my eventual choosing of a title that entered into conversation with his own marvellous book *Crow Country*, and he has been a great critic, a superb writing friend and an inspiration. Jonathan Darlington, possibly fed up with years of otter talk, encouraged me to take my study to Exeter University, where I met two of my best critics, Andy Brown and Sam North. These two believed wholeheartedly in the book from the very start, and their enthusiasm and helpful criticism have been invaluable. Sam introduced me to my brilliant literary agent, Clare Conville, who has been more than wonderful. Clare's passion and vision have made all the difference, and I can't thank her enough for her part in helping to make this book happen. Clare's partner at Conville and Walsh, Patrick Walsh, also influenced the development of the book with some astute and enlightening questions in its early stages. Many other key people have been generous with their time: Jimmy Watt, in reading over sections of the *Ring of Bright Water* material, and Brian Keeble, in making useful suggestions about Kathleen Raine. It was a privilege to meet both. Anne Williamson helped greatly with my research on Henry Williamson, and the musician Harry Williamson, Henry's son, was also hugely helpful and has become a great friend. Listening to Harry's *Tarka Symphony*, originally composed for the film version of *Tarka the Otter*, has become a part of my routine. I can't thank James Williams enough for sharing his profound knowledge of otters, for his friendship and advice, for his forgiveness each time I put my foot in it, and especially for the many rousing and entertaining phone conversations. I can only hope my pronouncements on otters and any oversights that are bound to occur in a book such as this will not offend him too much, and I apologise for any of the latter that might have blundered into the text. My wonderful editor Bella Lacey and all the team at Granta have been especially lovely to work with. Benjamin Buchan, the eagle-eyed and brilliant copyeditor, was astute, flexible and good-humoured when I had all but had enough. Many others deserve my thanks: otter experts, academics, scientists, writers, naturalists and poets have been generous with their encouragement, advice and time. These are, in no particular order: Rob Strachan, whose depth of knowledge still astounds me; Neil and Barbara Taylor, who were the best friends anyone could wish for – the reserve at Denmark farm is a testament to their vision; John McMinn to whom I owe thanks, for putting me up at short notice and for pointing me in the right direction; and Jos Smith, for always knowing exactly the right thing to say in a crisis or for that matter at any other time; Tim Dee, for his encouragement, inspiration and advice. Thanks to Kenneth Steven, for his poetry and just for being who he is; to Robert Macfarlane, for responding so kindly to my questions and whose writing remains an inspiration; to Hugh Warwick, for being my humorous hedgehog counterpart and for some hilarious otter moments; to Trevor Beer, for some excellent inside info; to Wendy Smaridge, for helping in all sorts of ways; to William Hawkins and Cathy Dagg, for being there at the start; to

Alex Leith, for lending me his maps and his house; to Paul Hyland, for his wonderful poems; to Jim Perrin, for his clever mind, his company and his supportive listening; to Becky Gethin, for her poetry and friendship; to Kevin O'Hara, for freely sharing his great knowledge and expertise, and for his superb handshake; to Chris Waters, for his poetry; to Fran Southgate, for meeting up and giving me huge amounts of information at short notice; to Elizabeth Chadwick, for her help and her tireless work on otter research at Cardiff University; to Annie Dillard, Barry Lopez and David Abram, for their inspiring writings; and to Katrina Porteous, for her poetry and for being an inspiration and a good friend.

Finally, I want to thank my tolerant and long-suffering family. Benjamin and Jennifer have endured my absences, forgone outings that normal families have, and never complained about being involved in innumerable thinly disguised treks, trips and holidays to suspiciously watery locations. Special thanks go to Rick Smaridge, whose patience, encouragement and love have been boundless.

Permissions

Bibliography and Further Reading

Abram, David, *The Spell of the Sensuous* (Vintage Books, New York, 1997).

Abram, David, *Becoming Animal* (Pantheon, New York, 2010).

Allen, Daniel, *Otter* (Reaktion Books, London, 2010).

Austin, Mary, *The Land of Little Rain* (Dover Publications, New York, 1996).

Bate, Jonathan, *The Song of the Earth* (Picador, London, 2000).

Beer, Trevor, *Tarka Country Explored* (North Devon Books, Bideford, 2004).

Bishop, Elizabeth, *Complete Poems* (Chatto & Windus, London, 2004).

Botting, Douglas, *The Saga of Ring of Bright Water* (HarperCollins, London, 1993).

Brooks, Steve, *A Field Guide to the Dragonflies and Damselflies of Great*

Britain and Ireland (British Wildlife Publishing, Hook, 1997).

Brown, Jr., Tom *The Tracker* (Prentice Hall, New Jersey, 1978).

Cameron, L.C.R, *Otters and Otter-Hunting* (L. Upcott Gill, London, 1908).

Cocker, Mark, *Loneliness and Time* (Secker and Warburg Ltd, London, 1992).

Cocker, Mark, *Crow Country* (Jonathan Cape, London, 2007).

Chanin, Paul, *The Natural History of Otters* (Christopher Helm, Bromley, 1985).

Chatwin, Bruce, *The Songlines* (Jonathan Cape, London, 1987).

Clifford, Sue, and King, Angela, *England in Particular* (Hodder and Stoughton, London, 2006).

Daston, Lorraine and Gregg Mitman, *Thinking with Animals: New Perspectives on Anthropomorphism* (Columbia University Press, New York, 2005).

Deakin, Roger, *Waterlog* (Vintage, London, 1999).

Dillard, Annie, *Pilgrim at Tinker Creek* (Harpers Magazine Press, New York, 1974).

Dillard, Annie, *Teaching A Stone To Talk* (Harper and Row, New York, 1982).

du Maurier, Daphne, *Jamaica Inn* (Victor Gollancz, London, 1936).

du Maurier, Daphne, *Enchanted Cornwall* (Michael Joseph Ltd, London, 1989).

Ehrlich, Gretel, *The Future of Ice* (Pantheon, New York 2004).

Farley, Paul, and Michael Symmons Roberts, *Edgelands* (Jonathan Cape, London 2011).

Frere, Richard, *Maxwell's Ghost* (Victor Gollancz, London, 1976).

Gatty, Harold, *Finding Your Way Without Map or Compass* (Dover Books, 1999).

Grahame, Kenneth, *The Wind in the Willows* (Methuen, London, 1908).

Hallett, Alyson, 'Migrating Stones', *Resurgence Magazine*, 256, September 2009

Haraway, Donna, *The Companion Species Manifesto* (Prickly Paradigm Press, Chicago, 2003).

Heaney, Seamus, 'Kinship', from *North* (Faber and Faber, London, 1975).

Hughes, Ted, 'An Otter', from *Lupercal* (Faber and Faber, London, 1960).

Hughes, Ted, *Memorial address*, published in *Henry Williamson: The Man, the Writings – A Symposium* (T.J. Press, Padstow, 1977).

Hyland, Paul, *Otter: Lutra lutra on the Stour* (Common Ground, Shaftesbury, 2001).

Hyland, Paul, *Art of the Impossible: New and Selected Poems* (Bloodaxe Books, Tarset, 2004).

Jamie, Kathleen, *Findings* (Sort of Books, London, 2005).

Kelsey, Elin, *Saving Sea Otters: Stories of Survival* (Monterey Bay Aquarium Press, Monterey, 1999).

King, Angela, John Ottaway and Angela Potter, *The Declining Otter: A Guide to its Conservation* (Friends of the Earth Otter Campaign Chaffcombe, Somerset, 1976).

Kingsley, Charles, *The Water Babies* (1893; Oxford University Press, Oxford, 1995).

Kruuk, Hans, *Otters: Ecology, Behaviour and Conservation* (Oxford University Press, Oxford, 2006).

Leopold, Aldo, *A Sand County Almanac* (Oxford University Press, Oxford, New York, 1949).

Lister-Kaye, John, *The White Island* (Longman, London, 1972).

Lister-Kaye, John, *Song of the Rolling Earth* (Time Warner, London, 2003).

Lister-Kaye, John, *At the Water's Edge* (Canongate Books, Edinburgh, 2011).

Lopez, Barry, *Arctic Dreams* (Harvill Press, London, 1999).

Mabey, Richard, *The Perfumier and the Stinkhorn* (Profile Books, London, 2011).

Macfarlane, Robert, *The Wild Places* (Granta Books, London, 2007).

Mason, C.F. and Macdonald, S.M., *Otters: Ecology and Conservation* (Cambridge University Press, Cambridge, 1986).

Maxwell, Gavin, *The Ring of Bright Water Trilogy* (Penguin Books, London, 2001).

Oswald, Alice, *Dart* (Faber and Faber, London, 2002).

Perrin, Jim, 'Holy Things', *New Welsh Review*, 86, Winter 2009.

Porteous, Katrina, *The Lost Music* (Bloodaxe Books, Newcastle-upon-Tyne, 1996).

Porteous, Katrina, *The Blue Lonnen* (Jardine Press, Ipswich, 2007).

Raine, Kathleen, *On a Deserted Shore* (Dolmen Press, Dublin, and Hamish Hamilton, London, 1973).

Raine, Kathleen, *The Lion's Mouth* (Hamish Hamilton, London, 1977).

Raine, Kathleen, *Autobiographies* (Skoob Books, London, 1991).

Raine, Kathleen, *Collected Poems* (Golgonooza Press, Ipswich, 2008).

Rilke, Rainer Maria, *Archaic Torso of Apollo*, from *Neue Gedichte II* (New Poems, part two), Frankfurt, 1908

Sanders, Scott Russell, *A Conservationist Manifesto* (Indiana University Press, Bloomington, 2009).

Snyder, Gary, *The Practice of the Wild* (Shoemaker & Hoard, Washington DC, 1990).

Somerset Otter Group, *Recent Research into Somerset Otters* (Taunton, 2010).

Steinbeck, John, *Cannery Row* (Penguin Classics, London, 2000).

Steven, Kenneth, *The Missing Days* (Scottish Cultural Press, Aberdeen, 1995).

Thomas, Keith, *Man and the Natural World* (Penguin Books, Middlesex, 1984).

Thoreau, Henry David, *Walden; or, Life in the Woods* (1854; Castle Books, Edison, New Jersey, 1999).

Tregarthen, J.C., *The Life Story of an Otter* (Cornwall Editions, Fowey, 2005).

Turberville, George, 'The Otter's Oration', from *The Noble Arte of Venerie or Hunting* (1575).

Turnbull, William, *Recollections of an Otter Hunter* (1896; Decimus Publishing, Jarrow on Tyne, 1983).

Walton, Izaak, *The Compleat Angler* (1653; Oxford University Press, Oxford, 1982).

Wayre, Philip, *The River People* (Collins and Harvill Press, London, 1976).

Wayre, Philip, *The Private Life of the Otter* (Book Club Associates, London, 1979).

White, Gilbert, *The Natural History of Selborne*, (1788; Penguin Classics, London 1987).

Williams, James, *The Otter Among Us* (Tiercel Publishing, Wheathampstead, 2000).

Williams, James, *The Otter* (Merlin Unwin Books, Ludlow, 2010).

Williamson, Henry, *Tarka the Otter* (Penguin Books, London, 1937).

Williamson, Henry, *The Story of a Norfolk Farm* (Faber and Faber, London, 1936).

Williamson, Anne, *Henry Williamson, Tarka and the Last Romantic* (Sutton Publishing, Stroud, 1995).

Wordsworth, Dorothy, *The Grasmere and Alfoxden Journals* (Oxford World's Classics, 1991).

Index

Alfred the Great, King, 137
algal blooms, 119
Aln, river, 291
Alfoxden, 136
ammonia, 316–17
An Teallach, 5
anal glands, 26
Angel of the North, 306–7
ash trees, 77, 175
Asian short-clawed otters (*Aonyx cinerea*), 247
Avalon Marshes, 133, 141
avocets, 284

badger setts, 41, 144, 227, 332
badger skulls, 227
badgers, 26–7, 98–9, 144, 148, 220, 227–8
Barnes, 312
Barnstaple, 104–5
Bassenthwaite, 256
bat boxes, 318
Beadnell, 287, 291, 296, 299
Beam Weir, 105, 109
Beddington Mill, 320
beetle cases, 75, 119, 231
Beinn Dearg, 5
Beinn Ghobhlach, 5

Benin River, 37
bent grass, 286
Beowulf, 135
Berwick-upon-Tweed, 281
Bible, the, 264–5
Bideford, 104–6
Bishop, Elizabeth, 133–4
bison, 29
bitterns, 313
Black Mountains, 212
Blyth, 281
Blyth, river, 291
Bodmin, 162–4, 168, 175
bog myrtle, 7, 18
bog oak, 135
Booth Museum, 9
boreholes, 315
Boswell, James, 20
Botting, Douglas, 49
Bow Creek, 316
breast milk, 129
Bridgwater, 132
Bristol Channel, 104, 106, 108, 119, 132, 212
brochs, 48
Bronze Age, 166, 181
brown trout, 92, 175, 256, 320
Brown, Tom, 229

Brown Willy (Bron Wennyly), 164–6, 171, 173
Brunel, Isambard Kingdom, 159
Buckfast, 93
Buckfastleigh, 94
bullhead, 75, 92, 216
bumblebees, 223, 307, 318
Burrow Mump, 129–30
Burrowbridge, 137
butterflies, 9, 219, 223, 307
Buttern Hill, 173

Caledonian forest, 3, 300
Californian sea otters, *see* sea otters
Cambrian Mountains, 216
Camel, river, 163–4, 168
Camelford, 163
Camusfeàrna, *see* Sandaig
canals, 39, 121, 136, 154
canalisation, 10, 130, 316, 324
Cape clawless otters (*Aonyx capensis*), 37
Cardiff University Otter Project, 126, 139, 263–4, 270–1
Cardigan Bay, 216
carp, 92, 131
castlings, 76
cats, 70–1, 84, 128
Celtic crosses, 90
Celts, 85, 213
cetaceans, 27
Chadwick, Liz, 271–4
Chatwin, Bruce, 145
Cheddar Gorge, 124
Cheviot Hills, 280–1, 287, 293
Chiltern Hills, 314
chimpanzees, 9

cobles, 290
Cocker, river, 242, 255–6, 259
Cockermouth, 255–7, 260
Cockshut, river, 323–5
coke, 75
common crane, 313
coots, 101, 143, 321, 333
Copplestone, 105
Coquet, river, 291, 293
Cornmill Dragonfly Sanctuary, 316
crabs, 41, 182, 184, 249, 298, 305
Cranmere Pool, 80–2, 86, 91
Cresswell Pool, 264
Croydon, 319–21
Crummock Water, 256–7

damselflies 222–3
Darlington, C. D., 8
Dart, river, 74, 80–2, 86–8, 90, 92, 94, 97, 328–9, 333
Dartmeet, 90, 93
Dartmoor, 80–4, 86–92, 97, 99, 109, 157, 168
Darwin, Charles, 8
De Lank, river, 168, 171
Deen City Farm, 321
deer, 5, 18, 20, 22, 90, 93, 130
denes, 296
Denmark Farm, 208–12, 215, 217–24, 234–8
Derwent, river, 256
Devon dialect, 84
diatoms, 317
Digitalis, 35
Dillard, Annie, 71–2
dippers, 259
Dittisham, 329

Dodds, James, 290
dodos, 10
dog roses, 73
dolerite, 293, 297
Domesday Book, 319
dormice, 220
drag, 76
dragonflies, 218–19, 231, 317, 323
Druridge Bay, 292–4
Drury, Chris, 326
du Maurier, Daphne, 169–71
ducks, 297
Duddon, river, 244–5
dunes, 299–300

eagles, 18, 20, 226–7
East Atlantic Flyway, 241, 292
ecological succession, 149–50, 294
eels, 54, 69, 129, 172, 260, 305, 321, 337
Ehrlich, Gretel, 141
Eigg, 56
Elms Pond, 321
elvers, 86, 173
Environment Agency, 126, 207, 224, 270–1, 284, 313, 325
European Water Framework Directive, 317
Exe, river, 132
Exeter, 97, 104–5, 131–2

Farne Islands, 297–8
Ferry House, 20
First World War, 68, 83, 104
Fish in the Classroom, 320
fishermen, 289–91
fishing traps, 260, 263–4

Five Sisters of Kintail, 62
flood alleviation, 324–5
flukes (*Pseudamphistomum truncatum*), 139
Fowey, river, 168, 173–80
fox burrows, 41
fox cubs, 182
fox skulls, 227
foxes, 22, 73, 98–9, 171, 220, 227, 306, 330, 332
Frere, Richard, 37–8, 44
frog ponds, 318
frogs, 119, 142, 167, 172, 194, 211
frogspawn, 210
Fyke nets, 260, 264

Gairloch, 5
Gateshead, 306
Gateshead Millennium Bridge, 304
geese, 60–1, 241, 287, 292, 299
Georgeham, 83
Glastonbury Tor, 141
Glen Shiel, 62
Glenelg, 19, 45, 50, 57
Godrevy, 201
goldcrests, 70–1
Gormley, Antony, 306
Grahame, Kenneth, 118
The Wind in the Willows, 29, 117–18, 180
granite hedges, 166
grass-snake nests, 318
great auks, 10
Gretna, river 256

Hadrian's Wall, 279, 281, 294
Halfdene the Dane, 282

Hallett, Alyson, 85
Ham Wall, 140
Hamilton James, Charlie, 80, 206–7
Hayle, 182, 185, 188, 197
Hayle, river, 181–201
Hayle tidal barrier, 190
Hayle Towans, 189
Heaney, Seamus, 124, 135
hearing, 76
Heart of Reeds, 326–7
Hembury Woods, 93
herring, 289–91
Hexham, 282–3
High Wilhays, 81
Highland Clearances, 48
holloways, 97
horses, 224
Hughes, Ted, 8, 69, 85, 91–2, 107, 154, 211–12

Industrial Revolution, 213, 293
Inny, river, 168
Irish Sea, 243–4, 257
Isle of Athelney, 137
Isle of Lewis, 39
Isle of Skye, 18–20, 22, 25, 39, 56, 294

Jackson, Kurt, 198
Jefferies, Richard, 96
Jesmond Dene, 304
Johnson, Samuel, 20

kelp, 31
Keswick, 256
King's Nympton, 105
King's Otterer, the, 265

kingfishers, 200, 318, 321
Kingsley, Charles
    The Water Babies, 196–7
kittiwakes, 304
Knoydart, 56
Kylerhea, 22

Lakenheath, 313
Lampeter, 215
'Land of the Two Rivers, The', 84
Landseer, Edwin, 267
Lanthwaite, 257
Lapford, 105
Lawley, Sam, 185–6
Lea, river, 313–19, 321
Leagrave Common, 314
Leamouth, 317
Lelant, 189, 197
Leopold, Aldo, 135
Lerryn, river, 179
Lewes, 322–7
Liles, Geoff, 211
Limehouse, 316
Lindale, 260
Lindisfarne, 299
Linnaeus, 26
Liskeard, 159, 175
Lister-Kaye, John, 51
Little Loch Broom, 5
Living Landscapes, 285, 304
London, 312, 315, 318
Long Nanny, 288–9
Looe, 264
Lorton Valley, 256, 259
Lostwithiel, 178–9
Lundy Island, 108
Luton, 314–15

Lynher, river, 159, 168

McMinn, John, 243–51
marsh harriers, 284
mass spectrometry, 118
Maxwell, Gavin, 21–2, 35–8, 40,
    43–57, 60–2
    Raven Seek Thy Brother, 21, 52
    Ring of Bright Water, 3, 18–19, 21,
        35, 37–8, 43–4, 48–9, 51–2, 54,
        56, 60–1
    The Rocks Remain, 21, 60
Maxwell's otters
    Chahala, 35
    Edal, 21, 37–8, 44, 53–4, 59–60
    Mijbil, 35–6, 38, 40, 43, 49–51, 54
    Teko, 38
Medway, river, 327
Mendips, 124
mergansers, 102, 259
Mesolithic people, 124
midges, 4, 18–19, 22, 55–8
migratory birds, 165–6, 241, 280, 299
mink, 26, 28–9, 193, 226, 322–3, 327,
    334
    compared with otters, 100–1
Mionictis, 9
molehills, 100, 102–3
moles, 103, 180, 209
Monterey Bay, 33
moorhens, 143, 321, 333
Morecambe Bay, 244
moth strips, 318
Mount Plynlimon, 212
mud horses, 133
mullet, 324
muskrats, 29, 135

mussel shells, 56, 291
mustelids, 26–30, 127
My Halcyon River, 80

Nelson, Lord, 319
Neot, river, 176
Newbridge, 93
Newcastle, 281, 285–6, 304, 312
Newcastle Weekly Chronicle, 269
newts, 318
Niger Delta, 37
North American river otter (Lutra
    canadensis), 79
North Devon cattle, 105–6, 141, 329
North Downs, 319
North Sea, 71, 280, 286–7, 292, 303,
    307
Northumberland, 50
Nutkins, Terry, 37–8, 57

oak trees, 96, 144
O'Hara, Kevin, 283–4, 286, 291–6
Okement, river, 81–2, 92
Olympic Park, 314, 316–18
opencast mines, 293, 301
orca, 22, 31–2, 34, 228
Oswald, Alice, 25, 145
otor, 25
otter cubs
    calls, 146and maternal care, 80,
        127, 147, 182–5, 235–6, 295
    numbers, 78–9, 127
    orphaned, 84, 139, 186–7, 247–8
    and otter hunting, 261–2
    and post-mortems, 275
    spraint, 174
otter holts, 41–2, 76–7, 129, 175,

182–3, 251, 337
Denmark Farm, 210–11, 221–2, 232
river Lea, 314, 318
otter hounds, 76, 261–3, 265–9
otter hunting, 11, 75–6, 207, 261–8
otter paths, 76–7, 121, 189, 251, 288, 291, 333–4
otter spraint, 42, 75–8, 86, 115–16, 118–19, 166–7, 174, 193–5, 286
compared with mink, 101
and feeding habits, 75, 90–1, 119, 177, 337
and nursing mothers, 183
otter surveys, 126–7, 207–8, 224–5, 242, 255, 280, 292, 311–12, 327
Otter Trust, 11–12
otters (*Lutra lutra*)
baculum, 268
breeding, 78–9, 127, 147, 206–7
calls, 146–7, 200, 254
camouflage, 248
cannibalism among, 183
climbing, 77
compared with mink, 100–1
deaths, 233–4, 262–4
eyesight, 77–8
facial muscles, 254
feeding, 78, 90–3, 119, 141–2, 147, 174, 177, 183, 234, 253–4, 305–6
feet, 9, 116–17, 274
fur, 41, 100, 183, 226, 263
gait on land, 7
gestation period, 79
hearing, 22
jaws, 8–9

legal protection for, 11, 263–4
lifespan, 234
maternal care, 80, 127, 147, 182–5, 235–6, 295
mating, 78–80, 182
metabolism, 119
nasal cavity, 9
pelts, 225–6
road deaths, 126–7, 137, 185–6, 189–90, 270, 275, 311, 327
size, 27, 41, 247
skeletons, 9–10
skulls, 7–8, 227
sleeping, 41–2, 77
species, 26
swimming, 24–5, 117, 184, 334
tail, 9, 100, 268
teeth, 8–9, 275
territoriality, 27, 295
toes, 237
tracks, 101, 115–18
vertebrae, 9
weight, 41, 100
whiskers, 7, 153
otters, stuffed, 247, 268
Ottery, river, 168
Ouse, river, 10, 322, 324, 326
Ouseburn, river, 304
owls, 302–3
barn owls, 83, 103–4, 211
tawny owls, 145, 338
oystercatchers, 7, 198, 299

Pacific Ocean, 30, 32, 34
Padstow, 163
parasites, 128, 137, 139, 228, 272, 276
Parrett, river, 129–30

PBDEs, 128–9
peat mines, 123
pelts, 225–6
Penzance, 188–9
peregrine falcons, 11, 71, 94
Perrin, Jim, 106
pheasants, 110, 330
Picts, 279, 282
pike, 75, 175, 321
pine martens, 26–7
pinnipeds, 26–7
Plym, river, 159
Plymouth, 159
polecats, 26–7, 193, 226
Polegate, 327
pollution, 10–11, 71, 93, 98, 127–9,
    207, 264, 280, 311
    and Cornish mines, 157–61
    river Lea, 315–17
    river Wandle, 319–20
    and sea otters, 32, 128
Porteous, Katrina, 289–91
Porth Kidney Sands, 189
Postbridge, 88
post-mortems, 272–6
Poulter Park, 321
Prestwick Carr, 285
puffins, 19, 108

Quantocks, 124

rabbits, 28, 295, 300
Raine, Kathleen, 49–52
raised mires, 216, 285–6
Rannoch Moor, 3, 18
ravens, 169, 299
red kites, 219, 232

red squirrels, 258–9
reed beds, 294
reindeer, 29
Relubbus, 194
Rhum, 56
rhynes, 120–1, 130, 149
Rilke, Rainer Maria, 99
Ring of Bright Water (film), 46
Romans, 165, 213, 279, 282, 294
Rough Tor, 164, 166, 174
rowan trees, 44, 49, 51
Royal Albert Bridge, 159
RSPB, 191, 312–13, 326
rudd, 159
Ruskin, John, 319
Rye Meads, 316
ryegrass, 220–1

Sabrina, 213
Sage, The, 304, 307
St Cuthbert, 297, 307
St Erth, 188, 191
St Ives, 188, 197
salmon, 41, 90, 92–3, 173, 191, 231,
    245, 268, 293, 336
    river Camel, 164
    river Fowey, 173
    river Teifi, 216
    river Thames, 313
    river Tyne, 281–2, 286, 305
sand eels, 19
sand martins, 19
Sandaig (Camusfeàrna), 18, 20, 22, 35,
    37, 44–5, 47–9, 52–7, 60–2
    Gaelic name, 52–3
sandcastles, 76, 230, 251
Saxons, 213

Scots pines, 300
sea otters (*Enhydra lutris*), 30–5
    fur, 30–3
    jaws, 31
    maternal care, 184–5
    teeth, 31
    toxoplasmosis deaths, 128
    weak heart, 32–3
sea trout, 54, 93, 293
    river Camel, 164
    river Fowey, 173, 178–9
    river Teifi, 216
    river Tyne, 281–2, 286
    river Winterbourne, 325
sea urchins, 31
Seahouses, 288, 296, 298–9
seals, 18, 20, 22–3, 27, 198, 227, 305
Secret World wildlife sanctuary, 186
Severn, river, 212–14, 216
Shapwick, 122
Shapwick Heath, 140–1, 144, 149
Shared Earth Trust, 209–10, 212, 215,
    218, 221, 224
sharks, 22, 31, 34, 228
shrimps, 294
Skrjabingylus nasicola, 228
skulls, 226–7
slugs, 27, 71, 231
smell, sense of, 76
snails, 71, 78, 119
snares, 192, 264
snow buntings, 241, 287, 299
snow hares, 225–6, 242
snowdrops, 136
snuff, 319–20
Snyder, Gary, 107
Somerset Levels, 120–54

song thrushes, 71
Sound of Sleat, 18, 22, 57
South Downs, 325–6
South Hams, 328
Southgate, Fran, 322–3
sparrowhawks, 70
sphagnum moss, 47, 82, 134–5, 151,
    172
Stalking Wolf, 229
starlings, 143–4, 148, 152, 177, 299
Steinbeck, John, 148
stickle poles, 268
sticklebacks, 77, 86, 92, 231, 289, 320,
    323
stoats, 26, 28, 225, 242
Stoke Gabriel, 329
Strachan, Rob, 224–32
Stratford, 317
Sunderland, 281
sundews, 82, 150
Sussex Wetland Landscapes Project,
    323
swallows, 165
swan mussels, 130–1
Sweet, Ray, 144
sycamore trees, 288, 299

Tamar, river, 159
'Tarka country', 104–9
Tasmanian tigers, 10
Taunton, 132
Tavy, river, 159
Taw, river, 81–2, 84, 89–90, 92,
    104–5, 108
Teifi, river, 208, 215–17, 222, 229
Teign, river, 81–2, 92
Thames, river, 312–14, 317–19, 321

Thames 21 charity, 320
Thesiger, Wilfred, 35
thistles, 211–12
Thoreau, Henry David, 53, 71, 135–6, 139
tidal flaps, 323–5
tin and copper mining, 157–8, 161, 188
toads, 167, 194, 209, 211
Tone, river, 129–30, 138
Torridge, river, 84, 104, 106, 108
Torridon, 4
Torridonian sandstone, 294
tors, 81
Totnes, 329
Tottenham Marshes, 316
toxoplasmosis, 128
tracking, 229–30
trackways, 144–5
Trago Mills, 175–7
Travers, Bill, 46
Trescowe Common, 195
Tughall Mill, 289
Turberville, George, 265
Turnbull, William, 268–9
Tweed, river, 291, 293
Tyne, river, 280–1, 283, 286, 291, 304, 307

Umberleigh, 105

veilwort (*Pallavicinia lyellii*), 150–1
Victorian art, 267–8
Vikings, 282, 287

Waddon Ponds, 320
Wadebridge, 163

Walker, 305
Walthamstow Marshes, 316
Walton, Izaak, 266, 319
Wandle, river, 318–21
Wandsworth, 319, 321
Wansbeck, river, 291
Water for Wildlife, 312
water voles, 29, 101, 129, 317–18, 321, 323, 331
watercress, 315
Watermeads, 321
Waterworks Nature Reserve, 316
Watt, Jimmy, 21–2, 37, 44, 59–62
Wayre, Philip and Jeanne, 11–12
Wear, river, 291
weasels, 26, 28, 225–8
West Penwith, 181
Westhay, 140, 144, 149, 151
wetland loss, rate of, 323
wetland restoration, 284–5, 326
Whin Sill, 286, 294
White, Gilbert, 165
whooper swans, 241, 287, 299
wild boar, 63–4
wildcats, 18
wildflower meadows, 318
Wildlife Trusts, 131, 283, 294, 312, 322, 326
Wilfrid, Bishop of York, 282
Williams, Elizabeth, 126
Williams, James, 125–30, 137, 140, 148, 158–9, 243, 246, 270
  The Otter Among Us, 119–20, 132
  and otter hounds, 266–7
Williamson, Henry, 67–70, 79, 82–6, 104, 146, 207, 226
  A Chronicle of Ancient Sunlight, 104

and otter hunts, 261–2, 268
Tarka the Otter, 10, 18, 67–9,
    79–80, 85, 89–91, 104–7,
    261–2, 268
wind turbines, 250–1
Windsor, 312
Winterbourne, river, 323–5
wolverines, 29–30
woodlice, 119
woodpeckers, 219, 232, 321
Wordsworth, Dorothy, 136
Wordsworth, William, 136, 244
Workington, 257
Wye, river, 212

Yeoford, 105
Yorkshire Dales, 29
Yoxon, Paul, 294
Ystwyth, river, 212